Springer Theses

Recognizing Outstanding Ph.D. Research

Aims and Scope

The series "Springer Theses" brings together a selection of the very best Ph.D. theses from around the world and across the physical sciences. Nominated and endorsed by two recognized specialists, each published volume has been selected for its scientific excellence and the high impact of its contents for the pertinent field of research. For greater accessibility to non-specialists, the published versions include an extended introduction, as well as a foreword by the student's supervisor explaining the special relevance of the work for the field. As a whole, the series will provide a valuable resource both for newcomers to the research fields described, and for other scientists seeking detailed background information on special questions. Finally, it provides an accredited documentation of the valuable contributions made by today's younger generation of scientists.

Theses are accepted into the series by invited nomination only and must fulfill all of the following criteria

- They must be written in good English.
- The topic should fall within the confines of Chemistry, Physics, Earth Sciences, Engineering and related interdisciplinary fields such as Materials, Nanoscience, Chemical Engineering, Complex Systems and Biophysics.
- The work reported in the thesis must represent a significant scientific advance.
- If the thesis includes previously published material, permission to reproduce this must be gained from the respective copyright holder.
- They must have been examined and passed during the 12 months prior to nomination.
- Each thesis should include a foreword by the supervisor outlining the significance of its content.
- The theses should have a clearly defined structure including an introduction accessible to scientists not expert in that particular field.

More information about this series at http://www.springer.com/series/8790

Li-Ciao Hong

Super El Niño

Doctoral Thesis accepted by
National Taiwan University, Taipei, Taiwan

Springer

Author
Dr. Li-Ciao Hong
Research Center for Environmental Changes
Academia Sinica
Taipei
Taiwan

Supervisor
Prof. LinHo
Department of Atmospheric Sciences
National Taiwan University
Taipei
Taiwan

ISSN 2190-5053 ISSN 2190-5061 (electronic)
Springer Theses
ISBN 978-981-10-9183-4 ISBN 978-981-10-0527-5 (eBook)
DOI 10.1007/978-981-10-0527-5

Printed on acid-free paper

This Springer imprint is published by SpringerNature
The registered company is Springer Science+Business Media Singapore Pte Ltd.

Supervisor's Foreword

In her thesis, Hong recognized the distinctness of "super El Niños" although only three episodes of "super El Niño" (1972/1973, 1982/1983, 1997/1998) have been observed during the instrumental era. Hong identified features which are universal among super El Niños but rarely seen in a conventional El Niño. Hong showed these features are statistically significant.

Two preconditions were identified by Hong as the deciding factors to breed the super El Niño. Eighteen months before the super El Niño peak, the Walker cell starts to intensify thus stashing positive heat content anomalies in the extra-equatorial western Pacific. Such intensified Walker cell appears to be in defiance of ENSO condition, whereas the ENSO source region only presents a neutral/weak La Niña phase. In the subsequent winter, a Meridional Mode developed in the Northern Hemisphere which emerges at a position much equatorward than the usual pre-Niño southern lobe. As winter advances, the Hawaii low sprawls via the mutual enhancement of the wind–evaporation–SST (WES) feedback and the Sverdrup balance, leading to a broad area of low-level westerly wind anomalies. The wind anomalies further induce eastward current anomalies along the northern flank of the equator. The SST anomalies within a key box N-CEP (0–7.5°N, 155°E–170°W) change phase from cold to warm and set up the stage that once the N-CEP deep convection is flared up in spring, vigorous interaction between ENSO and the mid-latitude circulation in the Southern Hemisphere takes place, as Hong described.

Upon the eastern Australian coast, the seasonal modulation of local westerly jet provides a viable Rossby wave source. It further facilitates the self-intensification of a transverse circulation between the N-CEP convection and a high-pressure anomaly near the South Australia. This loop particularly adapts at driving the cross-equatorial flow near New Guinea that accounts for the unusual growth of a super El Niño. Called as the "Southern Hemisphere Booster" (SHB, Hong et al.

2014 GRL), this mechanism enlarges the Bjerknes instability framework with important implication that the super El Niño owes its strength to the aggressive intervention from mid/high latitude in both Hemispheres.

Dr. Hong has keen eyes and light speed efficiency. The casting of super El Niño should shed new light on future global warming related topic.

Taipei, Taiwan
December 2015

Prof. LinHo

Contents

Chapter 1
Introduction

1.1 Super El Niño

El Niño and Southern Oscillation (ENSO) is a coupled atmosphere-ocean phenomenon that not only dominates tropical climate variations but also exerts a broad influence worldwide. The diverse behavior of ENSO motivates researchers to classify this phenomenon into various types to study their impacts and mechanisms. For example, El Niño events have been classified according to frequency (e.g., Rasmusson et al. 1990), magnitude (e.g., Wolter and Timlin 1998; Stephens et al. 2007), onset time (Xu and Chan 2001; Horii and Hanawa 2004), or zonal structure of sea surface temperature (SST) (Ashok et al. 2007; Kao and Yu 2009; Kug et al. 2009). Not intending to add new classification, a set of El Niño events is singled out in this study based on a threshold of magnitude. Figure 1.1 shows that among 18 identified El Niño events during the instrumental era of 1952–2010, only three events (1972/1973, 1982/1983, 1997/1998) with Niño-3 SST anomalies attain values substantially above the rest by at least one standard deviation. These extreme events are referred to as "super El Niño" and the rest as "regular El Niño."

Despite a limited number of super El Niño events that have been recorded in the modern period, they account for a disproportionate share of the known features of El Niño due to their large magnitude and their uniform pattern. Therefore, the grouping of super El Niño provides an attractive starting point to study El Niño. This new class of super El Niño events raises several questions. Do super El Niño events possess enough common features to merit a special class, or are they merely a stochastic picking from normality? What are the factors that cause this runaway growth of El Niño? Why do they appear so rarely yet seemingly following a decadal periodicity? Are there any precursors or preconditions leading to a super El Niño? To answer some of the questions raised, three topics are discussed in this dissertation.

L.-C. Hong, *Super El Niño*, Springer Theses,
DOI 10.1007/978-981-10-0527-5_1

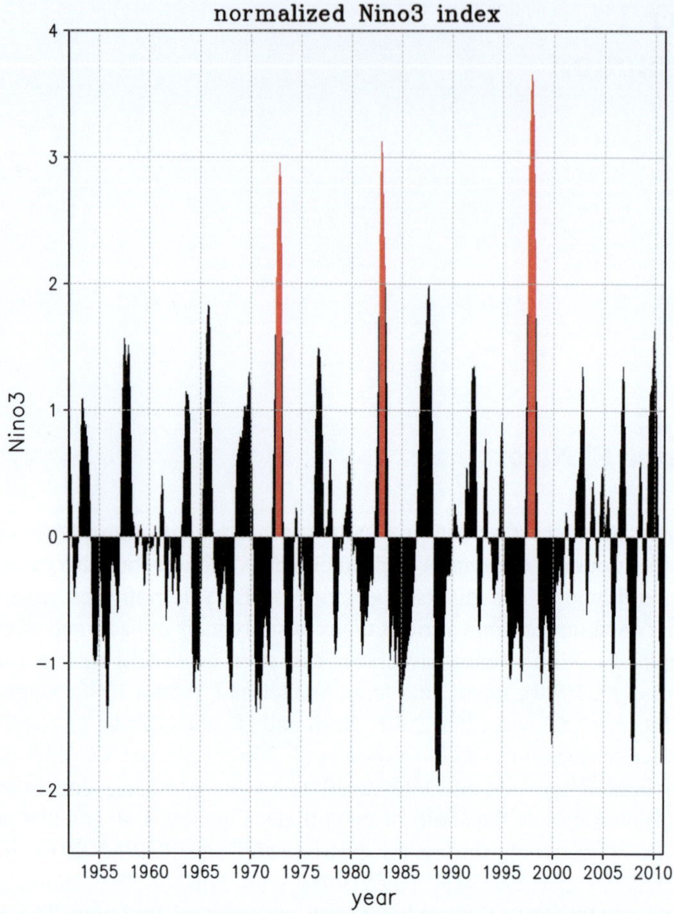

Fig. 1.1 Normalized Niño-3 index calculated using 6-month to 8-year band-pass-filtered anomalous SST data from 1952 January to 2010 December. Events with index values exceeding two standard deviations are plotted in *red* (Reprinted from Hong et al. 2014 with permission from John Wiley & Sons, Inc., New York, USA)

1.2 Topics and Literature Review

In the first topic, we ask, besides their large amplitude, can the super El Niño events be statistically singled out as an exclusive group? Are these events, other than identified by a threshold of conventional ENSO indices, exhibit structural similarities which are common among the super El Niño events, but almost unseen in the regular El Niño events? In other words, we are looking for the unique features of the super El Niños.

A K-means cluster analysis is performed to serve as a statistical basis and discussed in Chap. 3.1. Moreover, the unique and uniform spatial and temporal structure of super El Niño group in different stages are demonstrated by listing 12 El Niño events, super El Niño composite and regular El Niño composite covered in the ERA40 dataset in the period of 1960–2000 in Chap. 3.2. Several key features will be further elaborated in Chaps. 4 and 5, since they are tightly linked with the preconditions and the extra-amplification of super El Niño events.

The second topic involves the preconditions of super El Niño events. It is believed that a super El Niño event has much earlier, perhaps more vivid, signs preceded its onset. In past substantial studies have suggested possible mechanisms and precursors within and outside the tropical Pacific to trigger ENSO, most of the research considers ENSO/El Niño events as a whole group, and rarely addresses the precursors based on ENSO's magnitude (an exception can be seen in Stephens et al. 2007). Yet, these studies do provide some clues to the triggering of a super El Niño. In the tropical Pacific, a buildup of the upper ocean heat content anomalies along the equatorial Pacific from the extra-equatorial regions (Wyrtki 1985; Jin 1997; Meinen and McPhaden 2001) provides a favorable oceanic condition for the onset of ENSO events; intra-seasonal variability of westerly winds such as the Madden-Julian Oscillation (MJO, Madden and Julian 1994) and westerly wind bursts (Harrison and Vecchi 1997) in the western tropical Pacific are suggested as ENSO triggers (e.g., McPhaden, 1999; Kessler and Kleeman 2000; Lengaigne et al. 2004), which can be further intensified by the eastward extension of warm pool (Eisenman et al. 2005) or low-frequency zonal winds (Seiki and Takayabu 2007a, b; Kug et al. 2010). On the other hand, climate variability in the tropical Indian Ocean, North Pacific and even Southern Hemisphere may also facilitate the occurrence of ENSO events. For example, Izumo et al. (2010) suggested that the Indian Ocean Dipole (IOD, Saji et al. 1999) in the autumn of the year following an ENSO event modulates the intensity of the Walker cell, resulting in the subsequent ENSO event by a process similar to the advective-reflective oscillator (Picaut, 1997).Anderson (2004), Anderson and Maloney(2006) and Anderson et al. (2013) proposed that 12 months ahead of the peak time of El Niño, changes in trade winds due to a low SLP anomaly in the subtropical North Pacific recharge the equatorial Pacific subsurface heat content. Moreover, a stationary planetary wave called North Pacific Oscillation (NPO, Rogers 1981) is regarded as an important precursor to trigger ENSO following the so-called Seasonal Footprinting Mechanism (SFM) (Vimont et al. 2001, 2003). According to the SFM, anomalous SST patches, imprinted by the NPO on the North Pacific during boreal wintertime, not only persist when the NPO decays in late spring, but further migrate south to the tropics with the corresponding winds of the NPO via the Wind-Evaporation-SST feedback (WES feedback, Xie and Philander 1994; Vimont et al. 2009) in the summer and then excite oceanic Kelvin waves to trigger ENSO. However, not every NPO leads to ENSO (Park et al. 2012). Thus, another phenomenon, Pacific Meridional Mode (PMM, Chiang and Vimont 2004), a meridional band structure of anomalous SST and winds in the Pacific that resembles the evolution of the NPO in the SFM during spring, is suggested as a more reliable precursor of ENSO. Chang et al. (2007) showed that the PMM precedes the majority of ENSO events and acts as a conduit to influence ENSO from the North Pacific extra-tropics. Therefore the PMM potentially contribute to the onset of ENSO. Similarly, a new study (Zhang et al. 2014) identified an analogous pattern of the PMM in the South Pacific (SPMM) and suggested that SPMM may more directly trigger ENSO than the PMM because of its location.

As suggested by these published studies, in addition to the tropical Pacific, the precursors of super El Niño events are very likely to be found in the tropical Indian Ocean and the North Pacific. Thus, in the second part of this dissertation (Chap. 4), these areas are chosen as the target regions and the 12 El Niño events are listed as shown in Chap. 3.2 to contrast signatures which robustly and exclusively exist in every super El Niño event while absent in most of the regular El Niño events before their onset. The underlying mechanisms associated with these preconditions breeding a super El Niño are then discussed.

The last topic investigates the mechanisms to the unusual growth of a super El Niño. Unlike the literature of preconditions for a super El Niño, various hypotheses and theories have been proposed to explain the underlying causes of the extraordinary magnitude of some El Niño events. They include internal nonlinearities that lead to bursting growth of some El Niño (Timmermann et al. 2003), state-dependent (multiplicative) noise (Jin et al. 2007; Levine and Jin 2010) such as strong westerly wind events under an El Niño favorable condition (Lengaigne et al. 2004), and nonlinear response of convection to SST (Okumura and Deser 2010; Choi et al. 2013). However, the attentions thus far have been mostly focused on the processes within the tropical Pacific, while forces outside the tropics have been overlooked, which could play an indispensable role in pushing El Niño's growth over the hurdle.

Therefore, in Chap. 5, evidence for a tight relationship that exists between a super El Niño and certain Southern Hemispheric anomalous circulation during El Niño developing stage is presented. It is hypothesized that these Southern Hemispheric features act to boost El Niño's growth and turn it into a super El Niño. In other words, it requires the right amount of push through the positive feedback between equatorial Pacific and Southern Hemispheric winter subtropical circulation during El Niño developing phase in early boreal summer and autumn for an El Niño to grow into a super El Niño.

While the K-means cluster analysis and uniform spatial and temporal structures of super El Niño events shown in Chaps. 3–5, the limited number of super El Niño in the observation may still be a concern. Therefore, a preliminary analysis is conducted on simulations of a state-of-the-art coupled climate model with a dozen super El Niño events and shown in Chap. 6. Although the simulated super El Niño events are incapable to be utilized to evaluate the preconditions of super El Niño founded in the observation due to a biased temporal evolution, the result supports that the boosting mechanism we proposed using observation data also contributes significantly to the growth of model simulated super El Niño events.

The rest of this dissertation is organized as follows. Chapter 2 describes the data used in this study. Chapter 3 discusses the K-means cluster analysis and shows a super El Niño atlas that includes all significant and unique features. In Chap. 4, two preconditions of a super El Niño event are suggested. The first one is a "super-ma"

La Niña and another is the simultaneous emergence of two low SLP anomaly systems over the Philippine Sea and the Hawaii Island. The linkage between super El Niño onset and these two preconditions are investigated and summarized in this chapter. Chapter 5 focuses on the major unique features of super El Niño events that boost super El Niño's magnitude during the developing stage: the Southern Hemisphere transverse circulation. A preliminary result of super El Niño performances in the state-of-the-art global climate model is then evaluated and concluded in Chap. 6. In Chap. 7, results are summarized and future work arising from the previous studies is discussed.

References

Anderson, B. T., 2004: Investigation of a Large-Scale Mode of Ocean–Atmosphere Variability and Its Relation to Tropical Pacific Sea Surface Temperature Anomalies. *J. Clim.*, **17,** 4089-4098.

Anderson, B. T., and E. Maloney, 2006: Interannual Tropical Pacific Sea Surface Temperatures and Their Relation to Preceding Sea Level Pressures in the NCAR CCSM2. *J. Clim.*, **19,** 998-1012.

Anderson, B. T., R. C. Perez, and A. Karspeck, 2013: Triggering of El Niño onset through trade wind–induced charging of the equatorial Pacific. *Geophys. Res. Lett.*, **40,** 1212-1216.

Ashok, K., S. K. Behera, S. A. Rao, H. Weng, and T. Yamagata, 2007: El Nino Modoki and its possible teleconnection. *J. Geophys. Res.*, **112**.

Chang, P., and Coauthors, 2007: Pacific meridional mode and El Nino-southern oscillation. *Geophys. Res. Lett.*, **34**.

Chiang, J. C. H., and D. J. Vimont, 2004: Analogous Pacific and Atlantic meridional modes of tropical atmosphere-ocean variability. *J. Clim.*, **17,** 4143-4158.

Choi, J., S.-I. An, S.-W. Yeh, and J.-Y. Yu, 2013: ENSO-Like and ENSO-Induced Tropical Pacific Decadal Variability in CGCMs. *J. Clim.*, **26,** 1485-1501.

Eisenman, I., L. Yu, and E. Tziperman, 2005: Westerly Wind Bursts: ENSO's Tail Rather than the Dog? *J. Clim.*, **18,** 5224-5238.

Harrison, D. E., and G. A. Vecchi, 1997: Westerly Wind Events in the Tropical Pacific, 1986–95*. *J. Clim.*, **10,** 3131-3156.

Hong, L.-C., LinHo, and F.-F. Jin, 2014: A Southern Hemisphere booster of super El Niño. *Geophys. Res. Lett.*, 2014GL059370.

Horii, T., and K. Hanawa, 2004: A relationship between timing of El Niño onset and subsequent evolution. *Geophys. Res. Lett.*, **31,** L06304,.

Izumo, T., and Coauthors, 2010: Influence of the state of the Indian Ocean Dipole on the following year/'s El Nino. *Nature Geosci*, **3,** 168-172.

Jin, F.-F., 1997: An Equatorial Ocean Recharge Paradigm for ENSO. Part I: Conceptual Model. *J. Atmos. Sci.*, **54,** 811-829.

Jin, F.-F., L. Lin, A. Timmermann, and J. Zhao, 2007: Ensemble-mean dynamics of the ENSO recharge oscillator under state-dependent stochastic forcing. *Geophys. Res. Lett.*, **34**.

Kao, H.-Y., and J.-Y. Yu, 2009: Contrasting Eastern-Pacific and Central-Pacific Types of ENSO. *J. Clim.*, **22,** 615-632.

Kessler, W. S., and R. Kleeman, 2000: Rectification of the Madden–Julian Oscillation into the ENSO Cycle. *J. Clim.*, **13,** 3560-3575.

Kug, J.-S., F.-F. Jin, and S.-I. An, 2009: Two Types of El Niño Events: Cold Tongue El Niño and Warm Pool El Niño. *J. Clim.*, **22,** 1499-1515.

Kug, J.-S., K. P. Sooraj, T. Li, and F.-F. Jin, 2010: Precursors of the El Niño/La Niña onset and their interrelationship. *J. Geophys. Res.*, **115,** D05106.

Lengaigne, M., and Coauthors, 2004: Triggering of El Nino by westerly wind events in a coupled general circulation model. *Clim Dyn*, **23,** 601-620.

Levine, A. F. Z., and F.-F. Jin, 2010: Noise-Induced Instability in the ENSO Recharge Oscillator. *J. Atmos. Sci.*, **67,** 529-542.

Madden, R. A., and P. R. Julian, 1994: Observations of the 40–50-Day Tropical Oscillation—A Review. *Mon. Wea. Rev.*, **122,** 814-837.

McPhaden, M. J., 1999: Genesis and Evolution of the 1997-98 El Niño. *Science*, **283,** 950-954.

Meinen, C. S., and M. J. McPhaden, 2001: Interannual Variability in Warm Water Volume Transports in the Equatorial Pacific during 1993–99*. *J. Phys. Oceanogr.*, **31,** 1324-1345.

Okumura, Y. M., and C. Deser, 2010: Asymmetry in the Duration of El Nino and La Nina. *J. Clim.*, **23,** 5826-5843.

Park, J.-Y., S.-W. Yeh, J.-S. Kug, and J. Yoon, 2012: Favorable connections between seasonal footprinting mechanism and El Niño. *Clim Dyn*, 1-13.

Picaut, J., F. Masia, and Y. du Penhoat, 1997: An Advective-Reflective Conceptual Model for the Oscillatory Nature of the ENSO. *Science*, **277,** 663-666.

Rasmusson, E. M., X. Wang, and C. F. Ropelewski, 1990: The biennial component of ENSO variability. *Journal of Marine Systems*, **1,** 71-96.

Rogers, J. C., 1981: The North Pacific Oscillation. *Journal of Climatology*, **1,** 39-57.

Saji, N. H., B. N. Goswami, P. N. Vinayachandran, and T. Yamagata, 1999: A dipole mode in the tropical Indian Ocean. *Nature*, **401,** 360-363.

Seiki, A., and Y. N. Takayabu, 2007a: Westerly Wind Bursts and Their Relationship with Intraseasonal Variations and ENSO. Part I: Statistics. *Mon. Wea. Rev.*, **135,** 3325-3345.

Seiki, A., and Y. N. Takayabu, 2007b: Westerly Wind Bursts and Their Relationship with Intraseasonal Variations and ENSO. Part II: Energetics over the Western and Central Pacific. *Mon. Wea. Rev.*, **135,** 3346-3361.

Stephens, D. J., M. J. Meuleners, H. van Loon, M. H. Lamond, and N. P. Telcik, 2007: Differences in Atmospheric Circulation between the Development of Weak and Strong Warm Events in the Southern Oscillation. *J. Clim.*, **20,** 2191-2209.

Timmermann, A., F. F. Jin, and J. Abshagen, 2003: A nonlinear theory for El Nino bursting. *J. Atmos. Sci.*, **60,** 152-165.

Vimont, D. J., D. S. Battisti, and A. C. Hirst, 2001: Footprinting: A seasonal connection between the tropics and mid-latitudes. *Geophys. Res. Lett.*, **28,** 3923-3926.

Vimont, D. J., J. M. Wallace, and D. S. Battisti, 2003: The seasonal footprinting mechanism in the Pacific: Implications for ENSO. *J. Clim.*, **16,** 2668-2675.

Vimont, D. J., M. Alexander, and A. Fontaine, 2009: Midlatitude Excitation of Tropical Variability in the Pacific: The Role of Thermodynamic Coupling and Seasonality*. *J. Clim.*, **22,** 518-534.

Wolter, K., and M. S. Timlin, 1998: Measuring the strength of ENSO events: How does 1997/98 rank? *Weather*, **53,** 315-324.

Wyrtki, K.,1985: Water displacements in the Pacific and the genesis of El Nino cycles. *Journal of Geophysical Research: Oceans*, **90,** 7129-7132.

Xie, S.-P., and S. G. H. Philander, 1994: A coupled ocean-atmosphere model of relevance to the ITCZ in the eastern Pacific. *Tellus A*, **46,** 340-350.

Xu, J., and J. C. L. Chan, 2001: The Role of the Asian–Australian Monsoon System in the Onset Time of El Niño Events. *J. Clim.*, **14,** 418-433.

Zhang, H., A. Clement, and P. Di Nezio, 2014: The South Pacific Meridional Mode: A Mechanism for ENSO-like Variability. *J. Clim.*, **27,** 769-783.

Chapter 2
Data

In this study, the monthly atmospheric variables are from the 40-year European Centre for Median-Range Weather Forecasts (ECWMF) Re-Analysis (ERA-40) (Uppala et al 2005). The SST data is from the Hadley Centre Sea Ice and Sea Surface Temperature data set (HadISST) (Rayner et al. 2003), and the sea surface height (SSH), which is utilized as a proxy of upper-ocean heat content and thermocline depth, the ocean currents and the ocean temperature are from Simple Ocean Data Assimilation (SODA) version 2.0.2 (Carton et al. 2005). The ocean data assimilation product from the German Estimating the Circulation and Climate of the Ocean project (GECCO) (Köhl et al. 2006) is used for the oceanic mixed-layer heat budget analysis in Chap. 4. The GECCO serves as an appropriate dataset for ocean temperature budget analyses since no artificial internal sources and sinks of properties was adding when calculating the estimated state.

Monthly anomalies were obtained as departures from monthly mean climatology for the period of 1958–2001, then the Kolmogorov–Zurbenko (KZ) filter (Eskridge et al. 1997) was applied to remove high frequency (less than 6 months) and longer-than-ENSO scale (8 years) variations. The KZ filter gives iterative moving average that removes high frequency variation relative to the window size; the method cleanly separates various time scales of meteorological variables and has the same accuracy as the wavelet method. Hence, the band-pass filter (3 month window, two-iteration KZ filter minus 43 month window, two-iteration KZ filter) in particular preserves the peaks within ENSO time scale yet alleviates the abruptness of 1976/1977 climate change. The filtered data were highly correlated ($r > 0.9$) with those using other commonly employed methods, e.g., a 6 year high-pass filter based on successive application of running means centered at 25 and 37 months (Zhang et al. 1997). As a result, our conclusions presented here do not depend on our choice of a specific filter, and only analyses of KZ-filtered data are discussed.

Regular El Niño events (excluding the three super El Niño cases) are defined as Niño-3 SST anomalies exceeding 0.5 standard deviation for at least 3 months. The time span covered by ERA-40 includes nine regular El Niño events (1963/1964, 1965/1966, 1968/1969, 1969/1970, 1976/1977, 1986/1987, 1987/1988, 1991/1992, and 1994/1995). An El Niño onset year is denoted by suffix (0), Y (0), and the year follows an El Niño event is denoted as Y (−1). Likewise, months in and before an El Niño onset year are also denoted by suffix (0), and (−1), respectively. For

L.-C. Hong, *Super El Niño*, Springer Theses,
DOI 10.1007/978-981-10-0527-5_2

example, October–November in the year before an El Niño event is abbreviated as ON (−1) and October–November in an El Niño year is abbreviated as ON (0).

Since the number of super El Niño is limited in the observation, the output of a coupled model, GFDL-ESM2M, is also used to investigate super El Niño with much more (12) events. The GFDL-ESM2M model output was obtained from the Coupled Model Intercomparison Project phase 5 (CMIP5) (Taylor et al. 2011) data archive (http://cmip-pcmdi.llnl.gov/cmip5/index.html). Simulated super El Niño in the preindustrial experiment for a 500 year simulation period was analyzed and compared to regular super El Niño in the model.

References

Carton, J. A., B. S. Giese, and S. A. Grodsky, 2005: Sea level rise and the warming of the oceans in the Simple Ocean Data Assimilation (SODA) ocean reanalysis. *J. Geophys. Res.*, **110**.

Eskridge, R. E., J. Y. Ku, S. T. Rao, P. S. Porter, and I. G. Zurbenko, 1997: Separating Different Scales of Motion in Time Series of Meteorological Variables. *Bull. Amer. Meteor. Soc.*, **78,** 1473-1483.

Köhl, A., D. Dommenget, K. Ueyoshi, and D. Stammer, 2006: The global ECCO 1952 to 2001 ocean synthesis, ECCO Report No.40. http://www.ecco-group.org/report_series.htm.

Rayner, N. A., and Coauthors, 2003: Global analyses of sea surface temperature, sea ice, and night marine air temperature since the late nineteenth century. *J. Geophys. Res.*, **108,** 4407.

Taylor, K. E., R. J. Stouffer, and G. A. Meehl, 2011: An Overview of CMIP5 and the Experiment Design. *Bull. Amer. Meteor. Soc.*, **93,** 485-498.

Uppala, S. M., and Coauthors, 2005: The ERA-40 re-analysis. *Q. J. R. Meteorol. Soc.*, **131,** 2961-3012.

Zhang, Y., J. M. Wallace, and D. S. Battisti, 1997: ENSO-like interdecadal variability: 1900-93. *J. Clim.*, **10,** 1004-1020.

Chapter 3
How to Distinguish a Super El Niño?

How to identify a super El Niño case? At first glance, this issue seems a trivial one. Any proper ENSO measurement (e.g., Niño-3 SST anomaly index) would bring in cases with large values, provided enough samples are available, provided normality can stand. A common and important measurement of extreme El Niño is advocated by Vecchi and Harrison (2006) who regarded the eastern extension of warm pool and convection anomalies along the equator as the unique characteristic of extreme El Niño. Their definition appears quite comprehensive to analyze model output like CMIP3 or CMIP5 database. Hong et al. (2014) provided another prospect that the super El Niño events can be singled out as a separated group. Instead of ranking El Niño events, they chose a threshold to define super El Niño due to a large amplitude difference between super and regular El Niño events. By this threshold, the defined super El Niño develops in a fundamentally way different from the rest. Hong et al. (2014) stressed this point based on a *K*-means cluster analysis and some extensive investigations on the structural difference from case to case which flushes out what we called the unique signature of super El Niño. Therefore, the El Niño events can be divided into the super El Niño and regular (nonsuper) El Niño groups (without any intention to do a new classification since we have no say over the regular El Niño group).

In this chapter, a *K*-means cluster analysis serving as a statistical assessment of the grouping by this threshold is first presented. The unique signatures of super El Niño are then demonstrated by listing the 12 El Niño events, super El Niño composite and regular El Niño composite in the same page in Sect. 3.2.

3.1 Cluster Analysis

To provide a statistical basis for the new group of super El Niño, a *K*-means cluster analysis (MacQueen 1967) is utilizing. The method of the *K*-means clustering has been used across a broad spectrum of subjects in the atmospheric and ocean sciences (e.g., Nakamura et al. 2009; Johnson and Feldstein 2010; Freeman et al. 2012). It is designed to partition a given data set into *K* clusters that are constrained by minimizing the sum of intra-cluster Euclidean distance (point-to-cluster-centroid). The

L.-C. Hong, *Super El Niño*, Springer Theses,
DOI 10.1007/978-981-10-0527-5_3

number of clusters K, as a nonhierarchical clustering process, has to be assigned in advance. To determine the optimal K, Rousseeuw (1987) developed a "silhouette" graphic; for each K, there is a silhouette value for each point that measures the similarity within the cluster comparing to dissimilarity between clusters. The silhouette value for the ith point (Si) is defined as Si = (bi−ai)/max(ai, bi) where ai represents the average distance from i to the other points in the same cluster, and bi denotes the minimum average distance from i to points in the other cluster. A silhouette value covers the range from −1 to 1. When the mean silhouette value of a data set is close to one, it indicates the clusters are well separated. A negative value marks an overlap between the clusters, or possible misplacement of points in the wrong cluster. Therefore, by looking over the silhouette value, one can determine the best separation of a data set.

The K-means cluster analysis was performed on three vital ENSO variables (Niño-3 SST, equatorial thermocline slope and equatorial zonal wind stress) that constitutes the Bjerknes instability (Bjerknes 1969; Wyrtki 1975) (Fig. 3.1). Figure 3.1a shows the mean silhouette value as a function of cluster number K. The optimal cluster number comes at three, with $K = 2$ being the second highest and $K = 4$ falling considerably. Figure 3.1b–d illustrate scattering diagrams of the three vital ENSO variables, with clusters shaded by different colors. Figure 3.1c, as the best separation of the clusters, stands out with a cluster consisting of super El Niño events (red dots), which is well separated from the rest. This super El Niño cluster displays a remarkable two standard deviation distance from its neighboring cluster in green, which is exclusively comprised of regular El Niño events. It is interesting to note that the partition between the super El Niño cluster and the regular El Niño cluster appears more obvious than the partition between the La Niña cluster and the ENSO-neutral cluster. The latter two clusters become separable only when four clusters are allowed as in Fig. 3.1d, which give a much smaller silhouette value. Even in Fig. 3.1b where the El Niño cluster has distanced itself from the non-El Niño years, the separation of the two clusters cannot match the maximum silhouette at $K = 3$ that highlights the super El Niño grouping.

3.2 Unique Signatures of a Super El Niño

In this part, unique features tightly linked with the evolution of super El Niño events are first presented. Other unique signatures of super El Niño events will then be introduced one by one following the chronology of a super El Niño, from the "Pre-Niño to onset" period (Sect. 3.2.2) to the developing period (Sect. 3.2.3), and a brief summary is given to conclude this section.

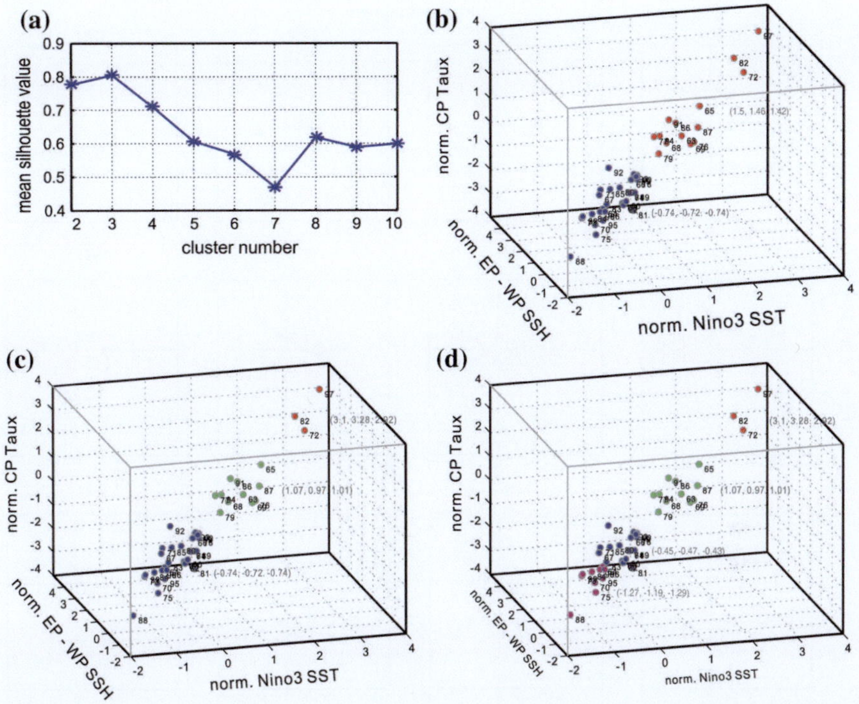

Fig. 3.1 **a** Mean silhouette value as a function of cluster number. **b–d** Scatter diagrams of normalized central equatorial Pacific zonal wind stress anomalies (averaged over 5°S–5°N, 160°E–140°W), Niño-3 SST anomalies (5°S–5°N, 90°W–150°W), eastern Pacific minus western Pacific SSH (2.5°S–2.5°N, 90°W–150°W minus 130°E–170°E) of October–December mean during 1959–2000. Different marks denote different clusters following the *K*-means cluster analysis. The *number* next to a circle denotes corresponding year; *the numbers in each parentheses* are the mean position of each cluster. Panels **b–d** have the cluster number *K* of 2–4, respectively (Reprinted from Hong et al. 2014 with permission from John Wiley & Sons, Inc., New York, USA)

3.2.1 Features Associated with the Evolution of a Super El Niño

3.2.1.1 Upper Ocean Heat Content in the Western Tropical Pacific

Figure 3.2 illustrates trajectories of 12 El Niño events in the phase space spanned by anomalous heat content averaged over the western north tropical Pacific (5°N–15°N, 130°E–180°) as abscissa and averaged over the western south tropical Pacific (5°S–15°S, 130°E–180°) as ordinate. The super El Niño events in the first column show a unique four-stage life cycle. Each stage (with the exception of the last stage) can be marked by a change point that is tightly locked to the annual cycle. During the first stage, super El Niño events have accumulated positive heat content

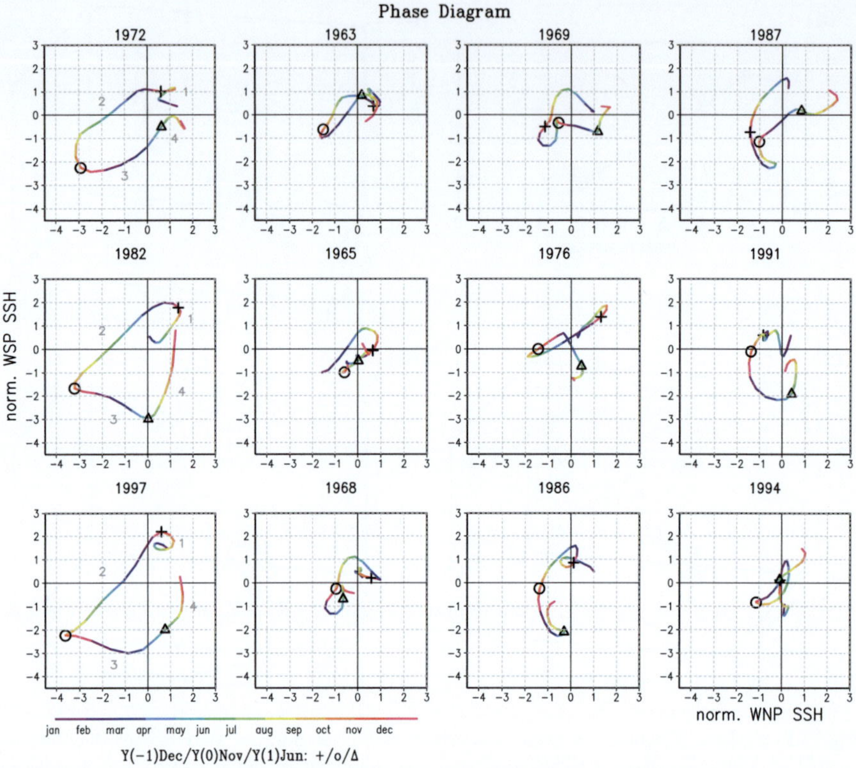

Fig. 3.2 Phase diagrams of normalized western north Pacific SSH anomalies (averaged for 5°N–15°N, 130°E–180°) and normalized western south Pacific SSH anomalies (5°S–15°S, 130°E–180°) for 12 El Niño events from January (−1) to January (1). *Cross*, *circle,* and *triangle* marks denote December (−1), November (0) and June (1), respectively. The numbers in the left column are referring to stages

anomalies in the western extra-equatorial Pacific for the whole Niño (−1) year and the value peaks in December (−1). Among regular El Niño events, only the 1976 El Niño case can stash the same amount of positive heat content anomalies in the western tropical Pacific, which lacks another necessary condition, as we will show later, to grow into a super El Niño. In the second stage from December (−1) to November (0), the super El Niño events rapidly and largely discharge the heat content in the western extra-equatorial Pacific. The discharge of the heat content is caused by the rapid development of a super El Niño, and the value reached minimum with an extraordinary intensity compared with other El Niño events in November (0). After November (0), the mature and decay phase of an El Niño, western north Pacific heat content anomalies increase from negative to neutral, but the recharge lags for 2–4 months in the south Pacific, leaving an arch structure in the third quadrant of phase diagrams. This implies that the negative heat content anomalies in the western south Pacific persistently provide cold subsurface water for the demise of a super El Niño, which results in a fast phase transition. Consequently, the phase diagrams of anomalous heat content in the tropical western north and south Pacific exhibit unique and internally uniform evolution for all super El Niño events.

3.2.1.2 Niño-3 SST Evolution

When examining the evolution of super El Niño events, a similar 7-year evolution in terms of Niño-3 SST anomaly index is surprisingly shown. Figure 3.3 shows normalized Niño-3 index for 12 El Niño events, which separated into 4 sets according to the temporal evolution of Niño-3 index, from January (−3) to January (4). The top figure consists of three super El Niño events. The evolution of Niño-3 index contains four stages. Two years before a super El Niño, there is usually a weak El Niño (Stage 1), and two successive weak La Niña events follows before the occurrence of super El Niño events (Stage 2). Stage 3 indicates that super El Niño events onset in spring, developing during summer and autumn, peaking in the end of the year, and have a sharp phase turning in the following year. The La Niña follows a super El Niño usually lingers and lasts for three years (Stage 4). Figure 3.3 also shows that three super El Niño evolve almost in the same pace in the second and the third stages. Thus, super El Niño at least offers a 3–5-year predictability if we can detect the precursors of super El Niño events two years before they onset. Even though some regular El Niño events have a similar temporal evolution as super El Niño in some stages, regular El Niño events provide little predictability since we even cannot be sure of the phase of ENSO in the next year of regular El Niño.

3.2.2 Pre-Niño to Onset Stage

3.2.2.1 Super-Ma La Niña

How early will a super El Niño sign surface? Figure 3.2 suggests that the positive heat content is expected to build up for the whole Niño (−1) year in the extra-equatorial areas. Nevertheless Fig. 3.3 also shows that the strong La Niña events like in 1965, 1976 Niño (−1) year (generally belong to the TBO type, with strong amplitude and the maximum cold SST anomaly appearing in the central Pacific) also precludes the emergence of a super El Niño (reasons we will give in Chap. 4). Therefore a mild type of La Niña appears to be the ideal breeder of the super El Niño.

On the other side, Fig. 3.3 clearly shows that the Niño-3 index could not uniquely determine the coming super El Niño. A sequence of mild La Niña in the Niño (−2) and Niño (−1) years provides the necessary condition, but not the sufficient condition for super El Niño onset. It turns out that strong, but not too

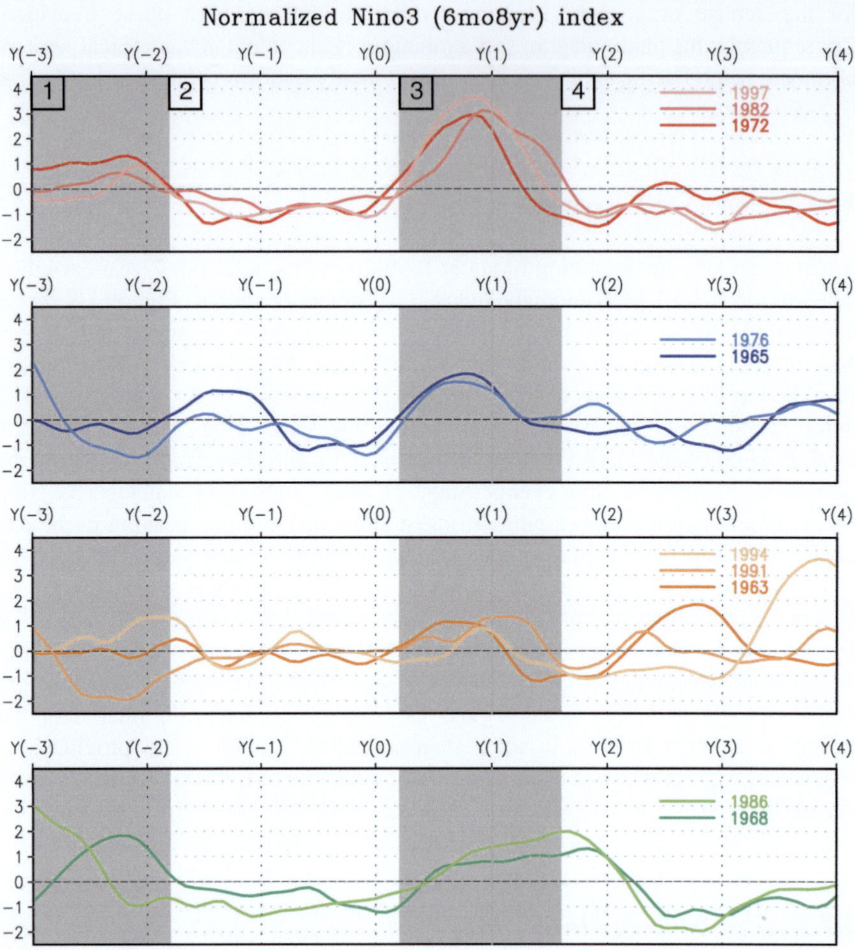

Fig. 3.3 Normalized Niño-3 index for 12 El Niño events from January (−3), denoted by Y (−3), to January (4), denoted by Y (4). These El Niño events are plotted in four different panels according to their temporal evolution. The 1969 and 1987 El Niño events are included in the same segments of the 1968 and 1986 El Niño events in the *bottom panel*. The number in the parentheses indicates January of the year relative to El Niño onset year. In the *top panel*, the number in the square denotes different stage in the 7-year evolution of a super El Niño

strong, Walker cell during Niño (−1) year marks the unique feature of a super El Niño, despite the strength of Walker cell is tightly related to the SST anomalies in the ENSO source area (as shown in Fig. 4.1n). Apparently, factors other than the La Niña accounts for the maintenance of the Walker cell prior to a super El Niño event. Some clues suggest contribution from the Indian Ocean SST anomalies (Izumo et al. 2010). It may be caused by indirect effect by prolonged tropospheric tropical temperature condition on the upper part of the equatorial Indian Ocean (Chiang and

Sobel 2002) or a monsoon anomaly. At this stage we can only give empirical evidences for what we call "super-ma" phenomenon.

3.2.2.2 Philippine Sea Low and Hawaii Low in D(−1)JFM(0)

During the boreal winter before a super El Niño event, the north Pacific reveals significant connections between extra-tropical forcing in the Northern Hemisphere and a super El Niño event. Figure 3.4 shows anomalous SLP fields in D(−1)JFM (0). Over the north Pacific, stationary waves dominate, yet super El Niño events commonly have a large-scale low SLP anomaly in the northern tropical Pacific. This large-scale low mainly consists of two low centers, one over the Hawaii Island and other located at the Philippine Sea. In 1997 D(−1)JFM(0), two centers are merged into one. This large-scale low is characterized by a relatively low-latitude position and its northern edge is nearly to the south of 30°N. In contrast, the

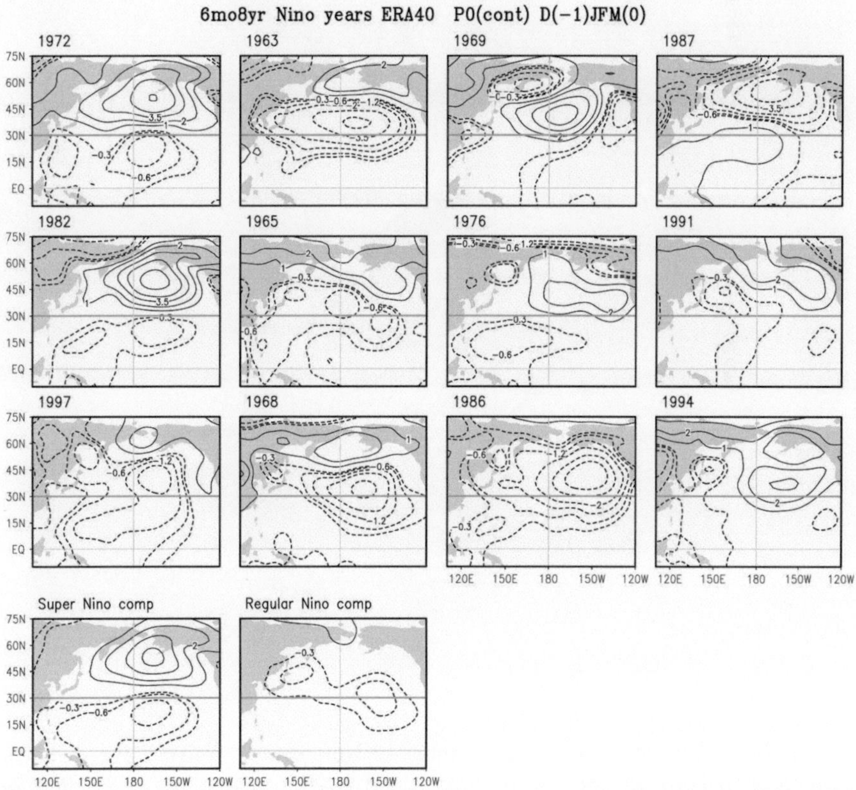

Fig. 3.4 Anomalous SLP (*contour* levels are −5, −3.5, −2, −1.2, −0.6, −0.3, 1, 2, 3.5, 5, and 7 hPa) for 12 El Niño events (*top three rows*), super El Niño composite (*bottom left*), and regular El Niño composite (*bottom right*) in D(−1)JFM(0)

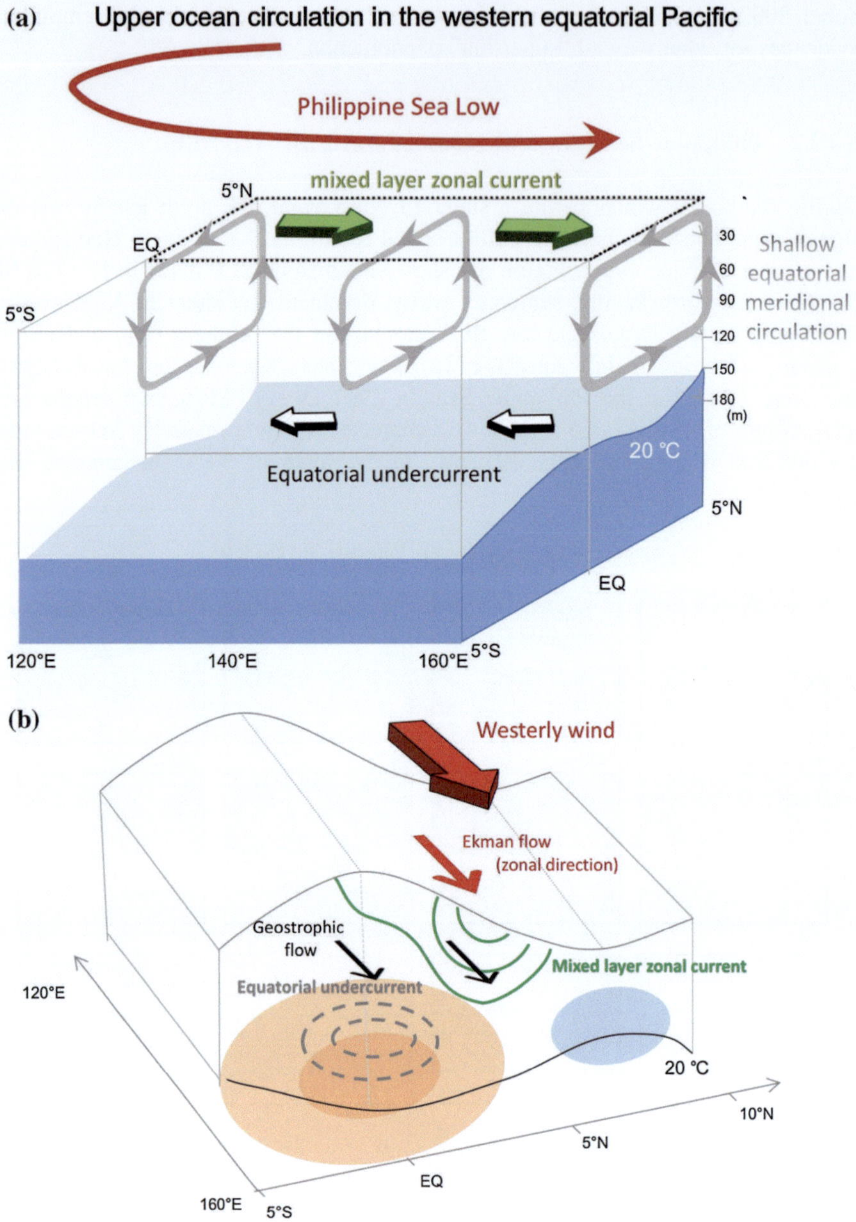

position of the low SLP anomaly in regular El Niño events, if they have, is much higher and the northern edge can reach to 50–60°N. This unique low SLP pattern in super El Niño contributes to its onset. Further discussions and its implications will be given in Chap. 4.2.

◀ **Fig. 3.5** Schematic diagram of (**a**) anomalous shallow vertical-meridional circulation and (**b**) the formation of anomalous mixed layer zonal currents in western equatorial Pacific during the spring of the super El Niño year. In (**a**), *red vector* denotes low-level westerly winds of the Philippine Sea low; *green arrows* denote the anomalous eastward currents in the mixed layer; *white arrows* denote anomalous westward equatorial undercurrents which move above the thermocline (20 °C); *gray loops* indicate a shallow vertical-meridional circulation between 0° and 5°N, from surface to 120 m along the western north Pacific. In (**b**), the *thick red arrow* denotes low-level westerly winds of the Philippine Sea low; the *red thin arrow* is Ekman flow in the zonal direction induced by winds; the *black thin arrows* are geostrophic flows maintained by a northward horizontal pressure gradient force induced by the upper ocean temperature anomaly (*shading*); *green contours* are eastward mixed layer zonal current anomaly and *dashed contours* are reduced equatorial undercurrent

3.2.2.3 Upper Ocean Circulation in Spring

From D(−1)J(0) to AM(0) of super El Niño years, anomalous currents in the upper ocean of the tropical Pacific are intimately linked to the Philippine Sea low and the Hawaii low. The western north Pacific upper ocean exhibits a shallow and small vertical-meridional circulation above the thermocline, as depicted in the schematic diagram (Fig. 3.5a) and shown in (Fig. 3.6, left column and Fig. 3.7). The anomalous low-level westerly winds of the Philippine Sea low induce equatorward Ekman flow from 4°N to the equator, which subsides at the equator, leaving the equator above the thermocline, and rises outside 4°N. Thus a shallow vertical-meridional oceanic circulation (gray loops) is observed under the anomalous westerly winds of the Philippine Sea low. According to the discharge oscillator (Jin 1997), the westerly winds and the associated shallow oceanic circulation implies that the equatorial water volume will discharge in the western Pacific in the following few months via Sverdrup transport. Although it provides no immediate benefit to develop an El Niño event, the coexisting strong surface eastward current in between the equator and 6°N along the western Pacific (green arrows in Fig. 3.5a and green contours in Fig. 3.5b) contributes the onset of a super El Niño, which is discussed in Chaps. 4.2–4.3. As shown in the schematic diagram Fig. 3.5b, this eastward mixed layer current is partly forced by the westerly winds of the Philippine Sea low (Ekman flow) and further maintained as a geostrophic flow by a northward horizontal pressure gradient force due to the subsurface warm water in the equator. In AM(0), this anomalous asymmetric (to the equator) eastward current penetrate to the eastern Pacific (Figs. 3.8 and 3.9). As Fig. 3.7 shown, the 1976 El Niño has a similar shallow circulation and strong eastward current to the north of the equator, but they are more symmetric to the equator in comparison to super El Niño events. This eastward geostrophic current becomes highly asymmetric to the equator after super El Niño onset due to an intensified cyclonic circulation over the western north Pacific in summer (JJAS, Fig. 3.6), which also induces a broad upwelling to the north of 6°N to cool the subsurface temperature. Thus eastward geostrophic current intensifies by the increased northward horizontal pressure gradient force.

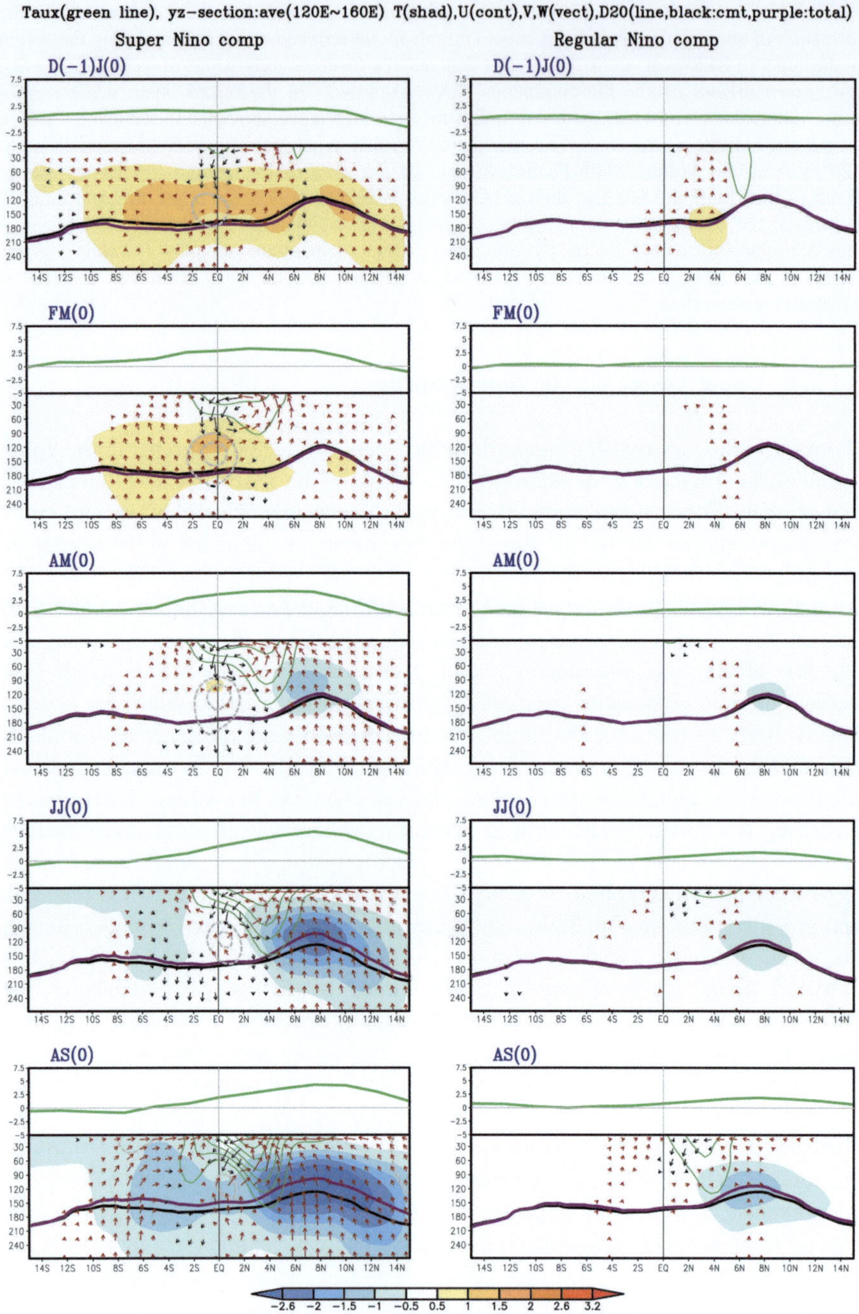
Taux(green line), yz-section:ave(120E~160E) T(shad),U(cont),V,W(vect),D20(line,black:cmt,purple:total)
Super Nino comp
Regular Nino comp
D(−1)J(0)
FM(0)
AM(0)
JJ(0)
AS(0)
−2.6 −2 −1.5 −1 −0.5 0.5 1 1.5 2 2.6 3.2

◀ **Fig. 3.6** In each panel, (*upper*) the anomalous zonal wind stress and (*lower*) latitude-height cross section of anomalous temperature (*shading*; °C), zonal current (*green* and *gray contours* with interval = 0.05 m/s), and meridional-vertical currents (*vector*) averaged in the western Pacific (120°E–160°E); the climatological and total temperature 20 °C isotherm are shown as *black* and *purple lines*, respectively. The vertical currents are multiplied by 1e4 for visual display. From the *top* to the *bottom panels* are bimonthly seasonal evolution, and (*left*) is for super El Niño composite and (*right*) is for regular El Niño composite

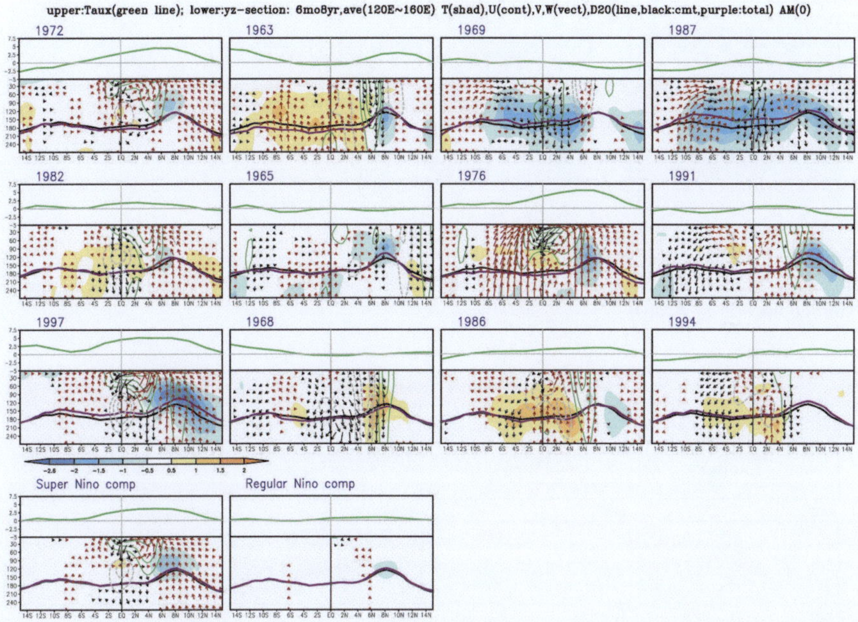

Fig. 3.7 Same as Fig. 3.6 but for 12 El Niño events (*top three rows*) and for super El Niño composite and regular El Niño composite (*bottom*) during AM(0)

In the central Pacific (Fig. 3.10), the anomalous eastward current in the mixed layer is relatively symmetric to equator, which emerges in FM(0). However, during the development of super El Niño (JJAS), low-level westerly winds are more intense along the north flank of the central tropical Pacific, and a vertical-meridional circulation, which is similar to the shallow circulation in the western north Pacific in FMAM(0) but much deeper, forms in the north-central equatorial Pacific. A noticeable large area downwelling in the whole eastern equatorial Pacific (Fig. 3.11) is induced by the reduction of the basin-wide trade wind, and this basin-scale trade wind reduction also decreases the equatorial undercurrent in the central and western Pacific.

Along the vertical-zonal cross section of the equatorial Pacific (Fig. 3.12), the mixed layer of the entire equatorial Pacific is dominated by a strong eastward current from FM(0), which is accompanied with a reduced equatorial undercurrent

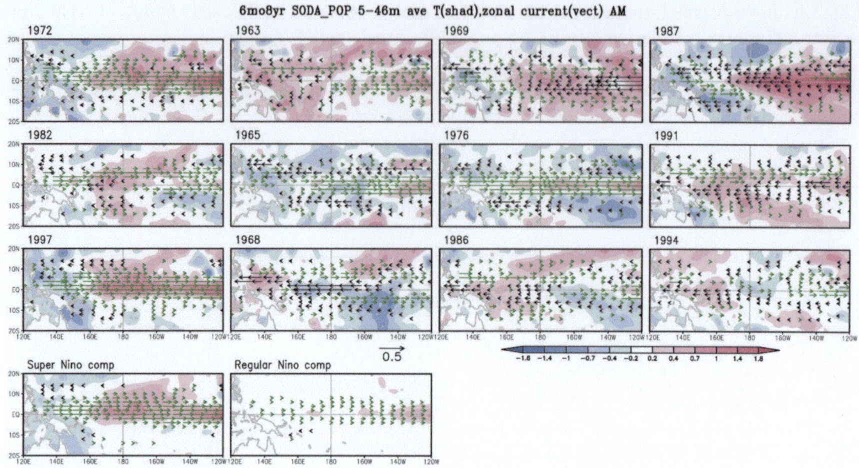

Fig. 3.8 5–46 m mean of ocean temperature (*shading*; °C) and zonal current (*vector*; m/s) anomalies using SODA POP data for 12 El Niño events, super El Niño composite and regular El Niño composite in AM(0)

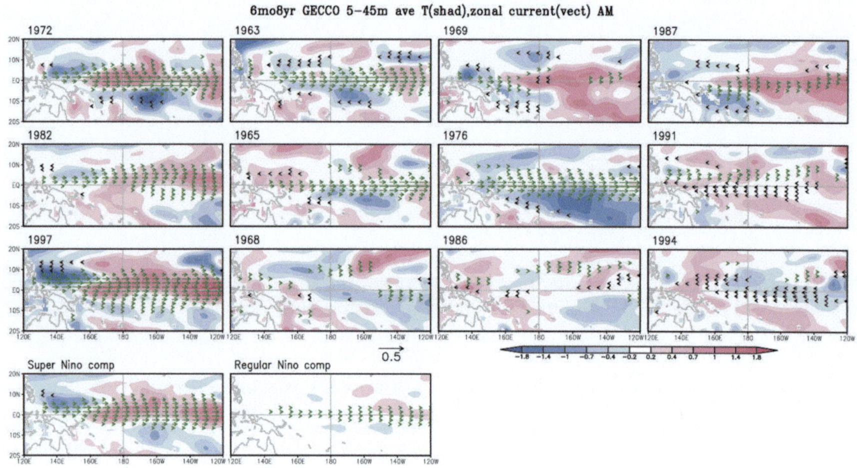

Fig. 3.9 Same as Fig. 3.8 except for GECCO data

mainly in the western Pacific sector as previously shown. Although the meridional-vertical cross section (Fig. 3.7) shows that three super El Niño events consistently reveal a relatively intense eastward current in the north of the equator, the equatorial eastward surface current in the 1982 super El Niño is not as obvious as the other two super El Niño events until summer (not shown). The strong eastward surface currents along the equator are also noticed by Santoso et al. (2013)

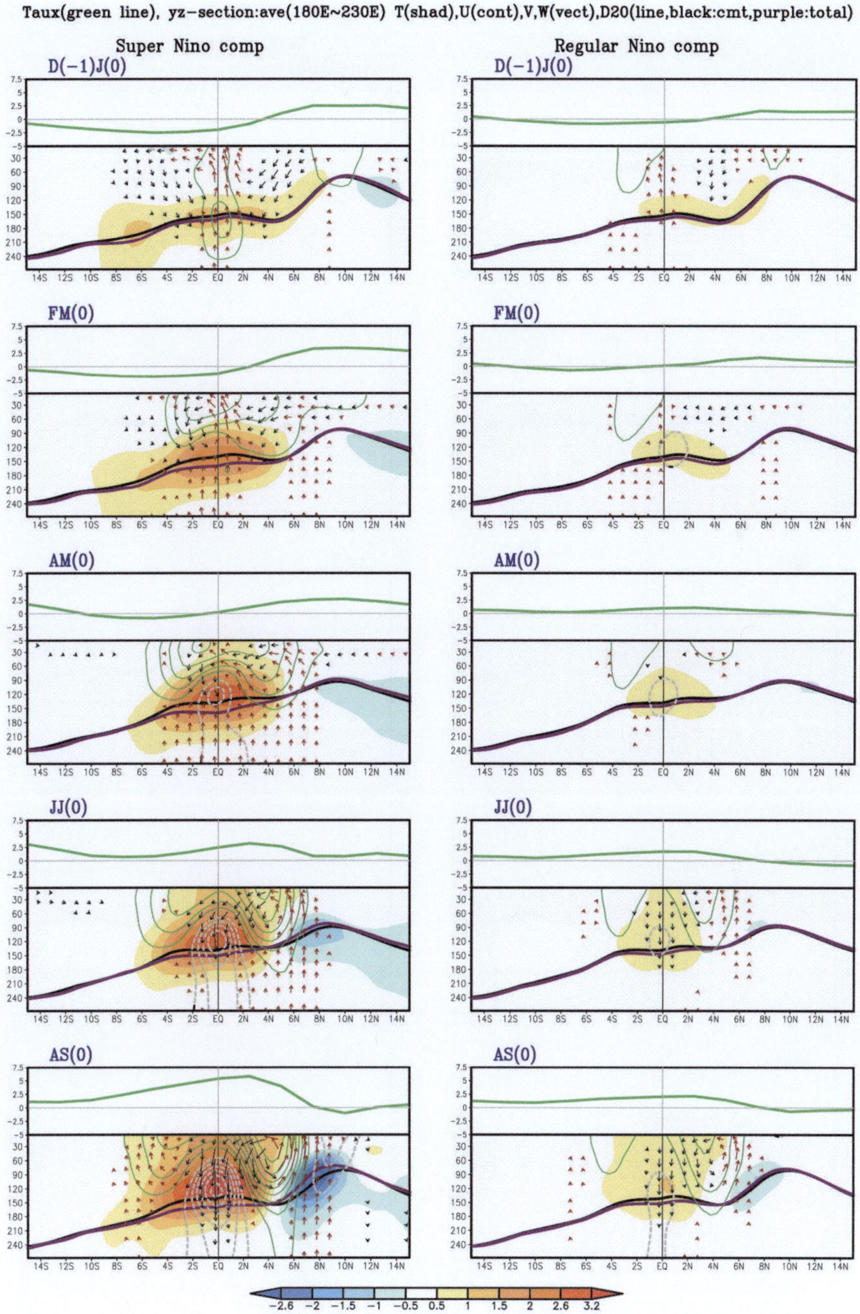

Fig. 3.10 Same as Fig. 3.6 except for the average over the central Pacific (130°W–180°)

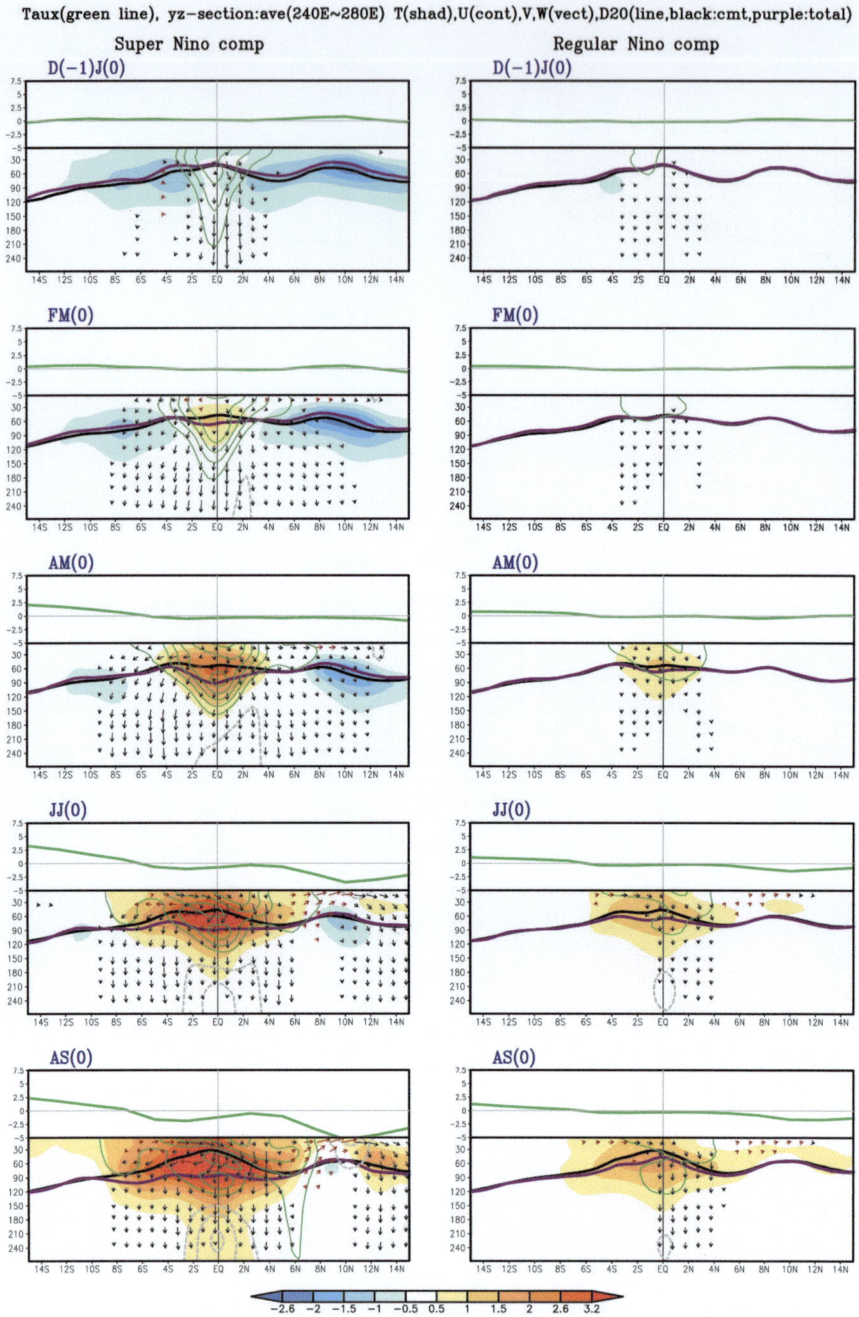

Fig. 3.11 Same as Fig. 3.6 except for the average over the eastern Pacific (80°W–120°W)

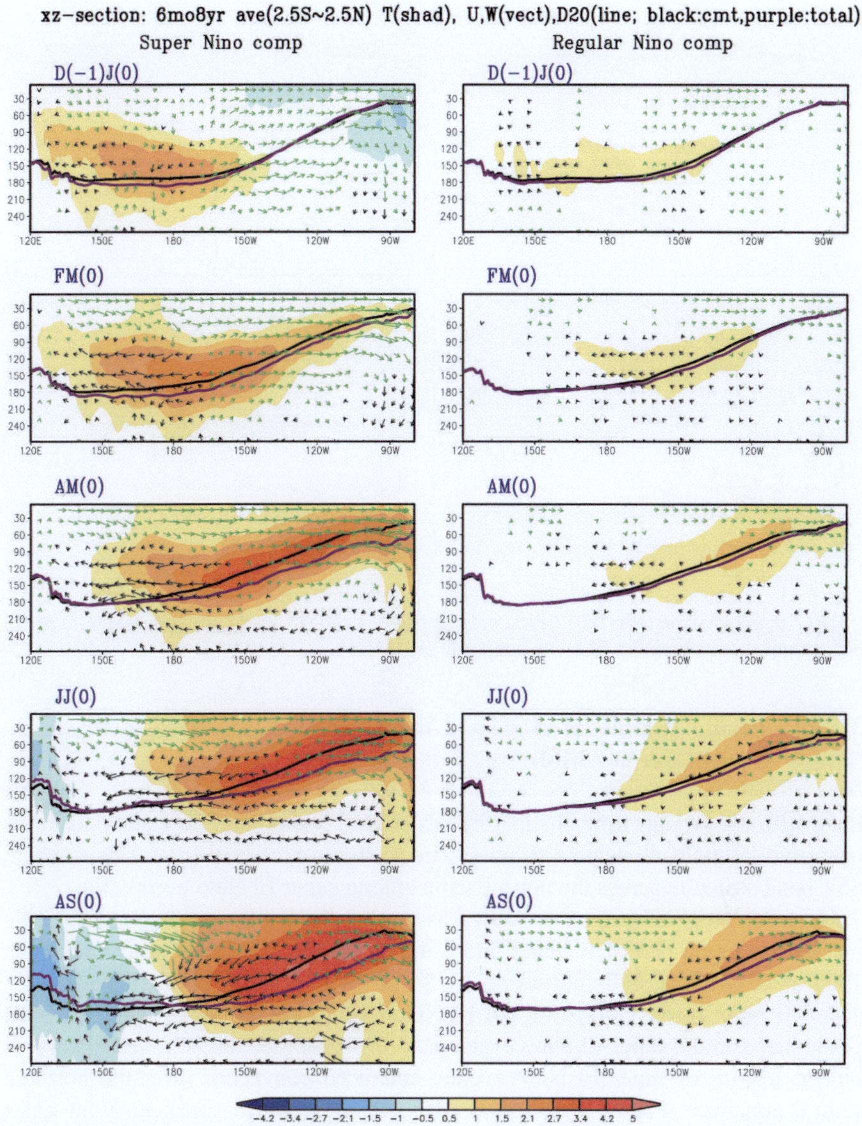

Fig. 3.12 Seasonal evolution of anomalous temperature (*shading*; °C), zonal-vertical current (*vector*; *green color* denotes eastward) along the equator (2.5°S–2.5°N) for super El Niño composite (*left*) and regular El Niño composite (*right*). The climatological and total temperature 20 °C isotherm are shown as *black* and *purple lines*, respectively. The vertical currents are multiplied by 1e4 for visual display

who regarded this phenomenon as one of the features of an extreme El Niño during its developing stage, which acts as a mechanism for extreme El Niños to propagate eastward.

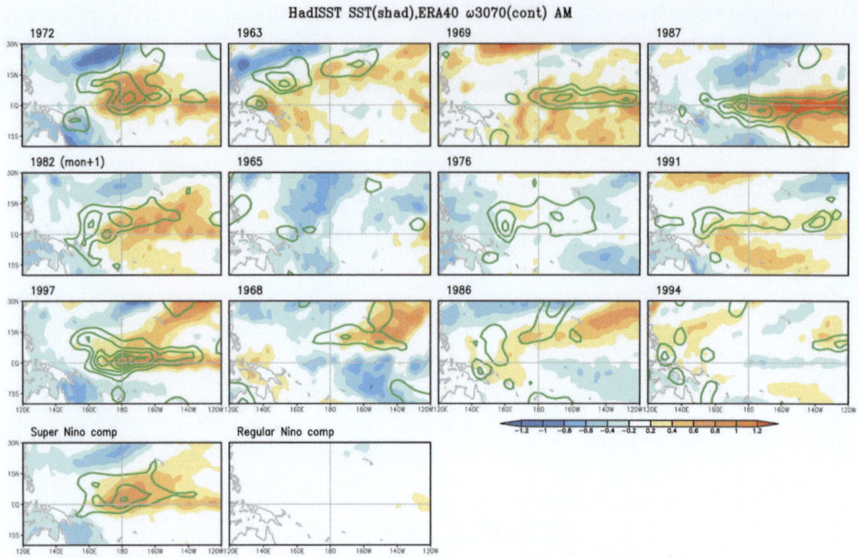

Fig. 3.13 Anomalous SST (*shading*; °C) and vertical pressure velocity averaged over 300–700 hPa (*contour*; interval is 0.012 Pa/s, only negative is shown) for 12 El Niño events, super El Niño composite and regular El Niño composite in AM(0)

3.2.2.4 Pacific Meridional Mode and the Associated Enhanced Convection in AM(0)

In AM(0), the average time of super El Niño onset, two oblique SST anomaly strips from mid-latitudes elongate southwestward to the tropics, and form a north-south SST band structure across the north Pacific among super El Niño events (Fig. 3.13). This anomalous SST pattern is known as one of the major features of the Pacific meridional mode (PMM, Chiang and Vimont 2004). The PMM also emerge in 1963 and 1986, but shifts northward by 5°–10° with relative to its counterparts in super El Niño cases. Comparing all El Niño events, the PMM signal is the most pronounced in the super El Niño events, and this phenomenon is linked to another unique feature of super El Niño events: enhanced convection over the northern central equatorial Pacific (0°–7.5°N, 155°E–170°W) when super El Niño onset (green contours in Fig. 3.13). Excluding two low-frequency El Niño (1969, 1987), the anomalous SST of the PMM during super El Niño events is particularly warm over the northern central equatorial Pacific compared to other El Niño events. Moreover, this warm SST is concurrent with enhanced convection within this area, which is proposed to trigger the extra-growth mechanism for super El Niño events. More details and discussions will be given in Chap. 4 when the onset of super El Niño is explored.

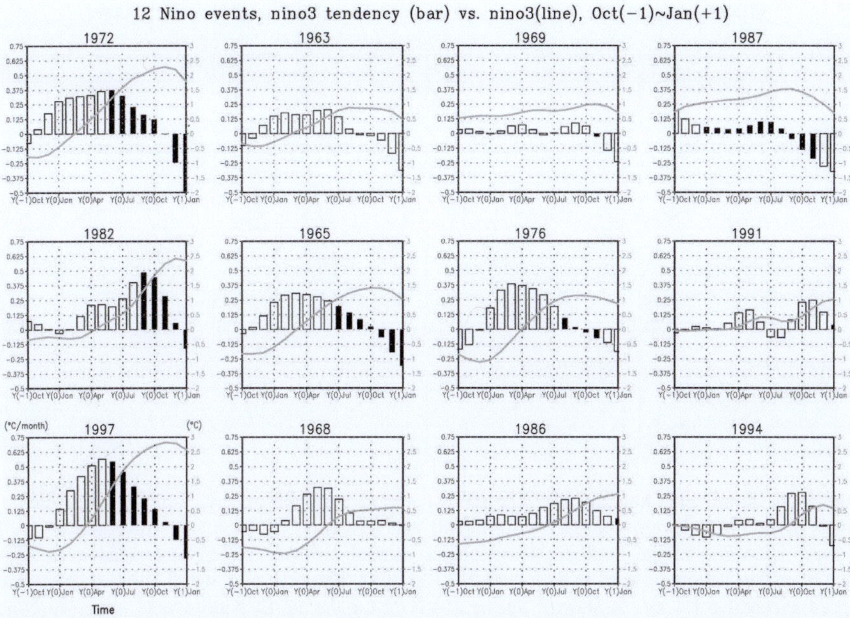

Fig. 3.14 Monthly Niño-3 SST index tendency (defined as Niño-3 index of the following month minus Niño-3 index of the preceded month; *bar*; °C/month; filled bar denotes Niño-3 value exceeding 1 °C) and monthly Niño-3 SST index (*line*; °C) segments for 12 El Niño events from Oct (−1) to Jan (1)

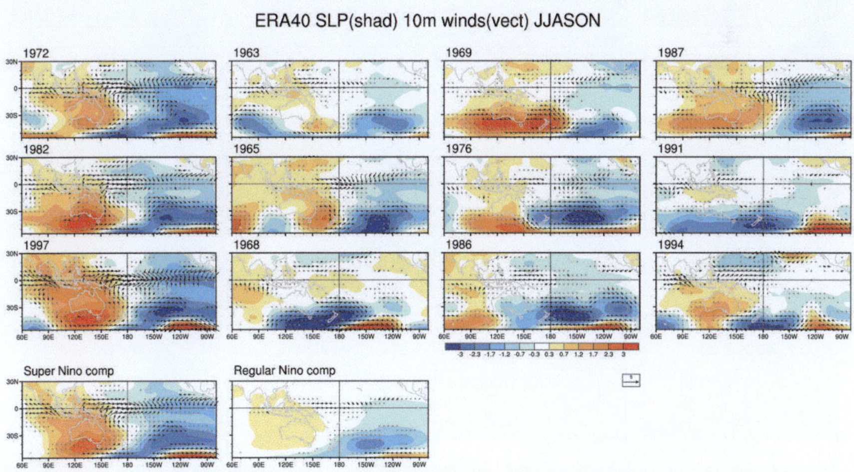

Fig. 3.15 Anomalous SLP (hPa; *shading*) and 10-m winds (only wind speed exceeding 0.5 m/s is shown, with *black arrows* denoting wind speed >1 m/s) during JJASON for the 12 El Niño events, super El Niño composite and regular El Niño composite (Reprinted from Hong et al. 2014 with permission from John Wiley & Sons, Inc., New York, USA)

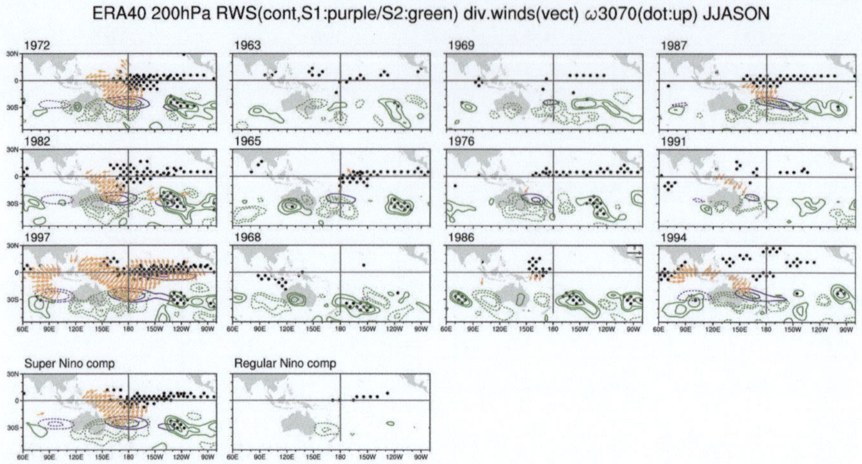

Fig. 3.16 Anomalous fields of 200 hPa RWS component of vortex stretching (*green contours*; interval: 1e-11 $1/s^2$, zero contours are omitted) and advection by anomalous divergence winds (*purple contours*) in the Southern Hemisphere; vertical pressure velocity averaged over 300–700 hPa (*dots* denote upward motion <-0.012 Pa/s); 200 hPa divergent winds (*vector*; only wind speed exceeding 1.4 m/s is shown) during JJASON for the 12 El Niño events, super El Niño composite and regular El Niño composite (Reprinted from Hong et al. 2014 with permission from John Wiley & Sons, Inc., New York, USA)

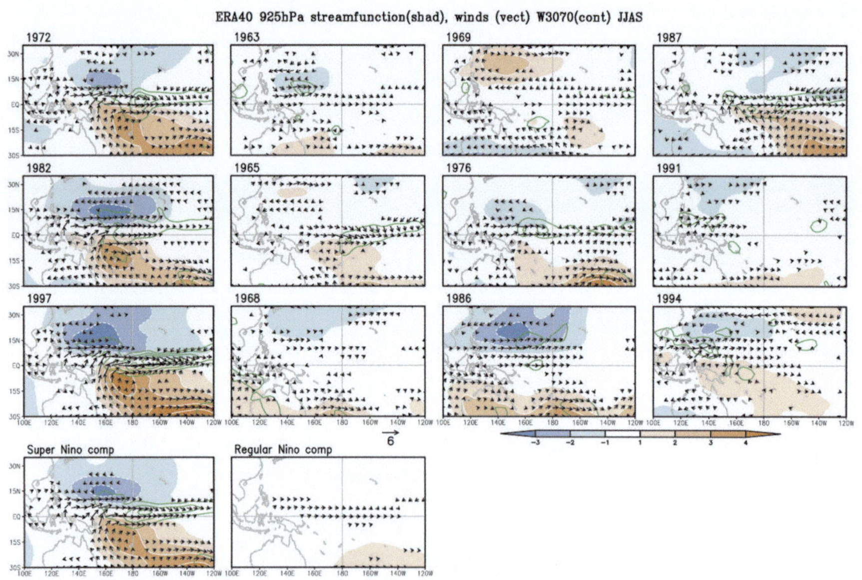

Fig. 3.17 Anomalous 925 hPa streamfunction (*shading*; 1×10^6 m^2/s), vertical pressure velocity averaged over 300–700 hPa (*contour*; interval is 0.012 Pa/s, only negative is shown) and 925 hPa winds (only wind speed exceeding 0.8 m/s is shown by *vector*) during JJAS for the 12 El Niño events, super El Niño composite and regular El Niño composite

3.2.3 Developing Stage

3.2.3.1 Niño-3 SST Tendency

Figure 3.14 shows the Niño-3 SST index tendency (bar) and Niño-3 index (line) for 12 El Niño events. Although the rapid intensification period in the 1982 super El Niño lags other two super El Niño by several months, the common feature of three super El Niño is that they still have large warming tendency when their Niño-3 index value reaches 1 °C during June to October/November. By contrast, the Niño-3 value of other El Niño events either fails to exceed 1 °C or has relatively weak warming tendency when Niño-3 attains 1 °C. This implies that other factors intensify amplifying mechanisms to defeat decaying mechanisms in super El Niño years during this period.

3.2.3.2 A Southern Hemisphere Transverse Cell

From June to November, the developing stage for ENSO, super El Niño events are distinct by a transverse cell which connects the central equatorial Pacific and the Southern Hemisphere. As shown in Fig. 3.15, a SLP dipole dominates the Southern Hemisphere mid-latitudes. Strong low-level southerly winds spun out from a high SLP anomaly over the southern Australia move northward and veer to right, forming a strong westerly wind belt in the entire western tropical Pacific. The low-level westerly winds converge in the central tropical Pacific, intensifying deep convection along the central and eastern tropical Pacific (dots in Fig. 3.16). The associated high-level divergent winds of the deep convection then flow westward to the Maritime Continent and also flow southwestward and converge over the northeast side of Australia. The induced subsidence over the eastern Australia finally returns to the low-level Australia high, constituting a transverse cell. The corresponding spatial pattern of the upper level anomalous Rossby wave source in the Southern Hemisphere among super El Niño events is tightly related with this transverse cell; thus displays similar structure (contours in Fig. 3.16). Moreover, the low-level circulation of the transverse cell in the western tropical Pacific is characterized by an asymmetric streamfunction dipole in the western Pacific due to the leaning of the southerly winds over east of New Guinea, and an intensified western north Pacific monsoon trough (Fig. 3.17). The maintenance of the Southern Hemisphere transverse cell and its effect on the super El Niño will be investigated in Chap. 5.

3.2.3.3 Teleconnection in the Southern Hemisphere

The existence of the Australian high arouses our interest in the impact of ENSO on the Southern Hemisphere. Here we show anomalous 500 hPa geopotential height

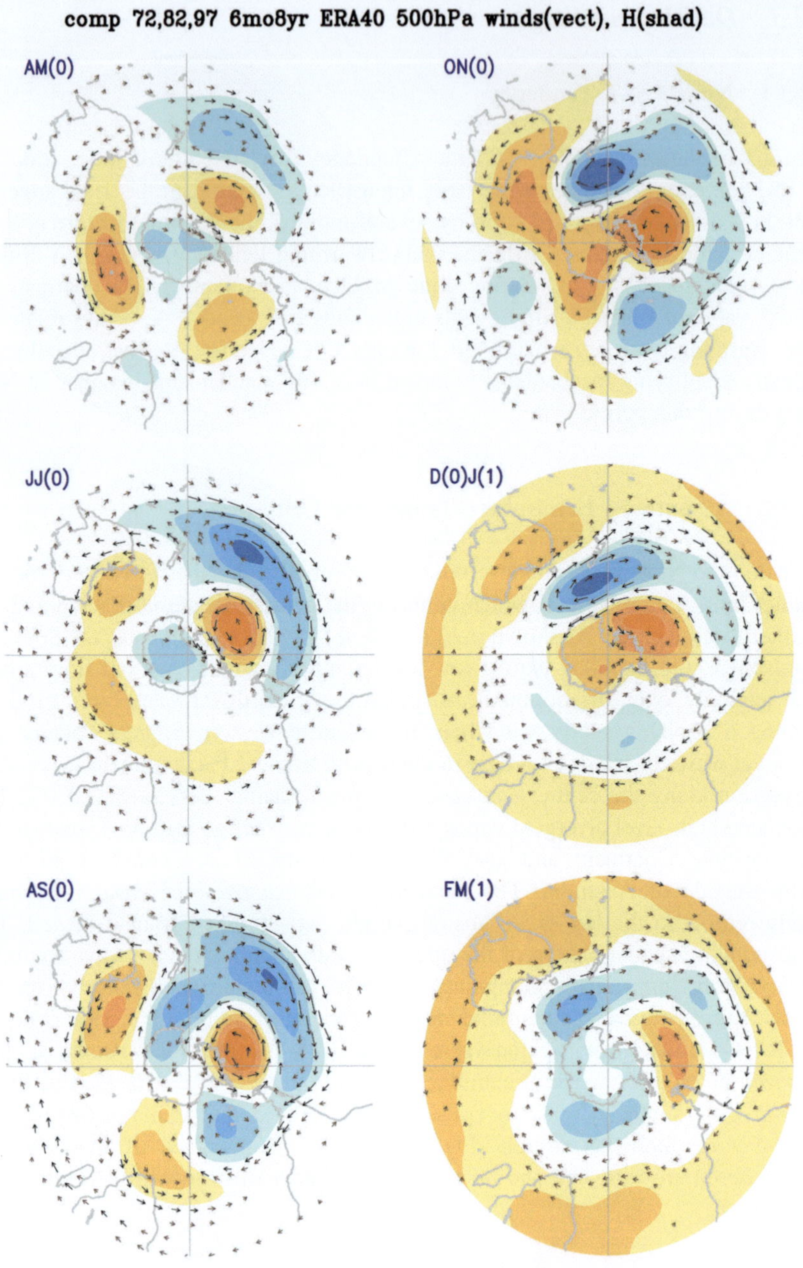

Fig. 3.18 Seasonal evolution of anomalous 500 hPa geopotential height (*shading*) and winds (*vector*) in the Southern Hemisphere for super El Niño composite

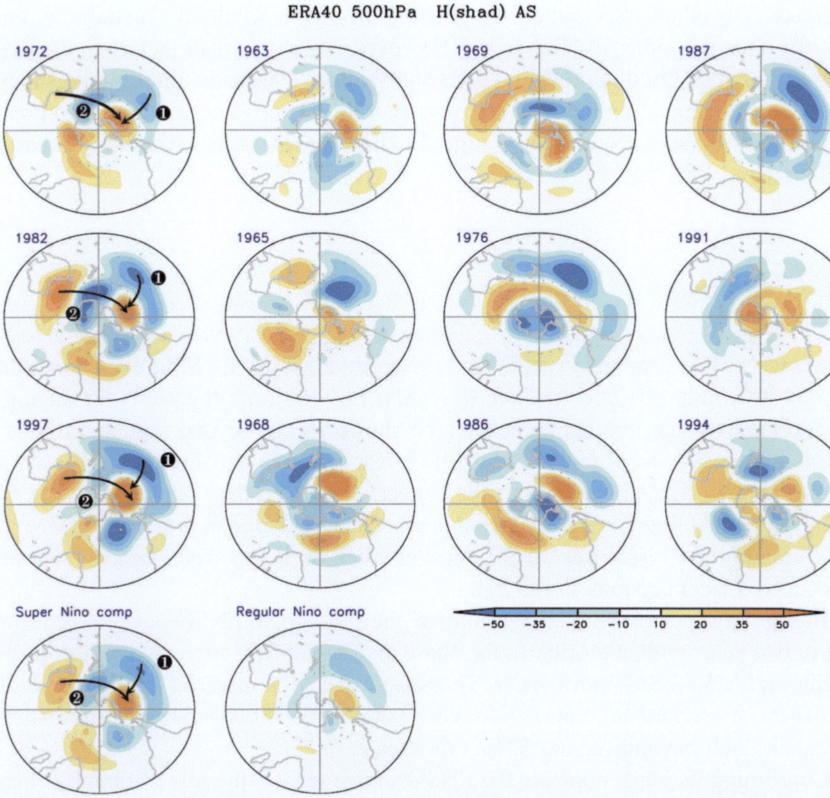

Fig. 3.19 Anomalous 500 hPa geopotential height (*shading*) field in the Southern Hemisphere for the 12 El Niño events, super El Niño composite and regular El Niño composite during AS(0). *Black arrows* indicate Rossby wave trains

fields to examine the teleconnection pattern related with super El Niño events. The influence of ENSO in the Southern Hemisphere (SH) are explored by many studies (e.g., Kwok and Comiso 2002; Park et al. 2004; L'Heureux and Thompson 2006). As Jin and Kirtman (2009) pointed out, teleconnections induced by ENSO lead ENSO and their correlation reaches maximum during austral winter due to local seasonality. Therefore, although several Rossby wave trains can be detected in super El Niño composite from AM(0) to FM(1) (Fig. 3.18), the uniform spatial pattern of super El Niño is the most robust in AS(0). As shown in Fig. 3.19, the common spatial structure of 500 hPa geopotential height anomaly is dominated by two Rossby wave trains of altering lows and highs in super El Niño events. They share the same node that is located to the south of South America, and each of the Rossby wave train almost spreads along the same path in three super El Niño events. These patterns are the so-called Pacific-South American patterns (Karoly 1989; Mo 2000). The PSA pattern emanating from the eastern tropical Pacific is well correlated with ENSO, so regular El Niño composite also shows this pattern.

Unlikely, the other PSA pattern originating from the southern Australia is not robustly shown in other El Niño event. However, other dominant pattern in the SH, such as Antarctic Oscillation (Thompson and Wallace 2000) and other forcings may modulate and generate different Rossby wave trains. Thus these waves will migrate in the following seasons, and three super El Niño events slightly lose the similarity in the second PSA pattern.

3.2.4 Summary

As shown in this chapter, in addition to magnitude, super El Niño events exhibit various structural difference relative to regular El Niño from the pre-Niño stage to its developing stage, though some may be shown in one or two regular El Niño. These unique signatures include a nearly 2-year La Niña situation before onset, a large-scale low SLP anomaly over the tropical Pacific in the winter before the El Niño year, which induces anomalous upper ocean circulation, a prominent PMM in the spring of the El Niño year; a SH transverse cell during its development, and the associated teleconnections in the SH.

Besides, a super El Niño also “controls” the subsequent ENSO development for one or two years with authority, as the phase diagram of SSH in the western tropical Pacific and Niño-3 SST index show. Thus a super El Niño marks at least a stretch of 3–5-years prescribed evolution, so each super El Niño provides forecasting potential much beyond regular ENSO timescale.

Concerning its impact outside the ENSO source region, the rule of thumb is that a stronger El Niño always exerts more influences on remote oceans and higher attitudes. This is indeed the case for the super El Niño. However, we only briefly discussed the teleconnection patterns in the SH; thus extensive investigations of this issue are suggested.

References

Bjerknes, J., 1969: Atmospheric teleconnections from the equatorial Pacific. *Mon. Wea. Rev.*, **97,** 163-172.

Chiang, J. C. H., and A. H. Sobel, 2002: Tropical tropospheric temperature variations caused by ENSO and their influence on the remote tropical climate. *J. Clim.*, **15,** 2616-2631.

Chiang, J. C. H., and D. J. Vimont, 2004: Analogous Pacific and Atlantic meridional modes of tropical atmosphere-ocean variability. *J. Clim.*, **17,** 4143-4158.Freeman, L. A., A. J. Miller, R. D. Norris, and J. E. Smith, 2012: Classification of remote Pacific coral reefs by physical oceanographic environment. *Journal of Geophysical Research: Oceans*, **117,** C02007.

Freeman, L. A., A. J. Miller, R. D. Norris, and J. E. Smith, 2012: Classification of remote Pacific coral reefs by physical oceanographic environment. *Journal of Geophysical Research: Oceans*, **117,** C02007.

Hong, L.-C., LinHo, and F.-F. Jin, 2014: A Southern Hemisphere booster of super El Niño. *Geophys. Res. Lett.*, 2014GL059370.

Izumo, T., and Coauthors, 2010: Influence of the state of the Indian Ocean Dipole on the following year/'s El Nino. *Nature Geosci*, **3,** 168-172.

Jin, D., and B. P. Kirtman, 2009: Why the Southern Hemisphere ENSO responses lead ENSO. *J. Geophys. Res.-Atmos.*, **114**.

Jin, F.-F., 1997: An Equatorial Ocean Recharge Paradigm for ENSO. Part I: Conceptual Model. *J. Atmos. Sci.*, **54,** 811-829.

Johnson, N. C., and S. B. Feldstein, 2010: The Continuum of North Pacific Sea Level Pressure Patterns: Intraseasonal, Interannual, and Interdecadal Variability. *J. Clim.*, **23,** 851-867.

Karoly, D. J., 1989: Southern Hemisphere Circulation Features Associated with El Niño-Southern Oscillation Events. *J. Clim.*, **2,** 1239-1252.

Kwok, R., and J. C. Comiso, 2002: Southern Ocean climate and sea ice anomalies associated with the Southern Oscillation. *J. Clim.*, **15,** 487-501.

L'Heureux, M. L., and D. W. J. Thompson, 2006: Observed relationships between the El Niño-Southern Oscillation and the extratropical zonal-mean circulation. *J. Clim.*, **19,** 276-287.

MacQueen, J. B., 1967: Some Methods for Classification and Analysis of MultiVariate Observations. *Proc. of the fifth Berkeley Symposium on Mathematical Statistics and Probability*, L. M. L. Cam, and J. Neyman, Eds., University of California Press, 281-297.

Mo, K. C., 2000: Relationships between low-frequency variability in the Southern Hemisphere and sea surface temperature anomalies. *J. Clim.*, **13,** 3599-3610.

Nakamura, J., U. Lall, Y. Kushnir, and S. J. Camargo, 2009: Classifying North Atlantic Tropical Cyclone Tracks by Mass Moments. *J. Clim.*, **22,** 5481-5494.

Park, Y. H., F. Roquet, and F. Vivier, 2004: Quasi-stationary ENSO wave signals versus the Antarctic Circumpolar Wave scenario. *Geophys. Res. Lett.*, **31**.

Rousseeuw, P. J., 1987: Silhouettes: A graphical aid to the interpretation and validation of cluster analysis. *Journal of Computational and Applied Mathematics*, **20,** 53-65.

Santoso, A., and Coauthors, 2013: Late-twentieth-century emergence of the El Nino propagation asymmetry and future projections. *Nature*, **504,** 126-130.

Thompson, D. W. J., and J. M. Wallace, 2000: Annular modes in the extratropical circulation. Part I: Month-to-month variability. *J. Clim.*, **13,** 1000-1016.

Vecchi, G. A., and D. E. Harrison, 2006: The termination of the 1997-98 El Nino. Part I: Mechanisms of oceanic change. *J. Clim.*, **19,** 2633-2646.

Wyrtki, K., 1975: El Niño—The Dynamic Response of the Equatorial Pacific Oceanto Atmospheric Forcing. *J. Phys. Oceanogr.*, **5,** 572-584.

Chapter 4
How to Trigger a Super El Niño?

In this chapter, we trace the earliest signals of super El Niño and discuss how they contribute to the onset of super El Niño. We suggest that a special type of La Niña in ON(−1) leads the peak time of super El Niño for at least 13 months, and external forcings from the North Pacific in D(−1)JFM(0) precondition the tropical Pacific for the super El Niño.

4.1 Precondition 1: Super-Ma La Niña

The year before the occurrence of a super El Niño, the tropical Indo-Pacific sector is dominated by a La Niña event. This is the type of La Niña to put two seemingly contradictory attributions under the same wing. The first is a well-organized enhanced Walker cell with anomalous easterlies in the western tropical Pacific that preconditions the ocean by generating large amount positive heat content anomalies stashed in the extra-equatorial region via Sverdrup transport. The second is to hold the central tropical Pacific SST anomalies in a nearly neutral state such that it would not pose a formidable obstacle when the La Niña has a chance to turn phase. Combining these two factors a "super-ma" La Niña (the mother of super El Niño) is seen to produce a super El Niño event next year.

Figures 4.1 and 4.2 illustrate these two characteristics of such a super-ma La Niña. Figure 4.1a–l lines up 12 El Niño cases of anomalous heat content, zonal wind stress during October–November in the year before El Niño [ON(−1)]. They are divided into three categories. The first column consists of the 1965, 1976 El Niño events. They have the strongest La Niña in Niño (−1) year among 12 events, as shown in Fig. 4.1n and Fig. 4.2. Their easterlies prevail in the central tropical Pacific and penetrate deep into the western side of 150°E. Moreover, the negative heat content anomalies are well developed reaching to the date line and the SST anomalies in the central Pacific drop to −1 °C, as later shown, which will set up an impassable hurdle for developing a super El Niño event. The third column of Fig. 4.1 covers all remaining seven regular El Niño events available by the ERA40 data. They appear more diverse than the other two columns. Some belong to the Niño (−1) year of a low-frequency El Niño episode and some store negative heat

L.-C. Hong, *Super El Niño*, Springer Theses,
DOI 10.1007/978-981-10-0527-5_4

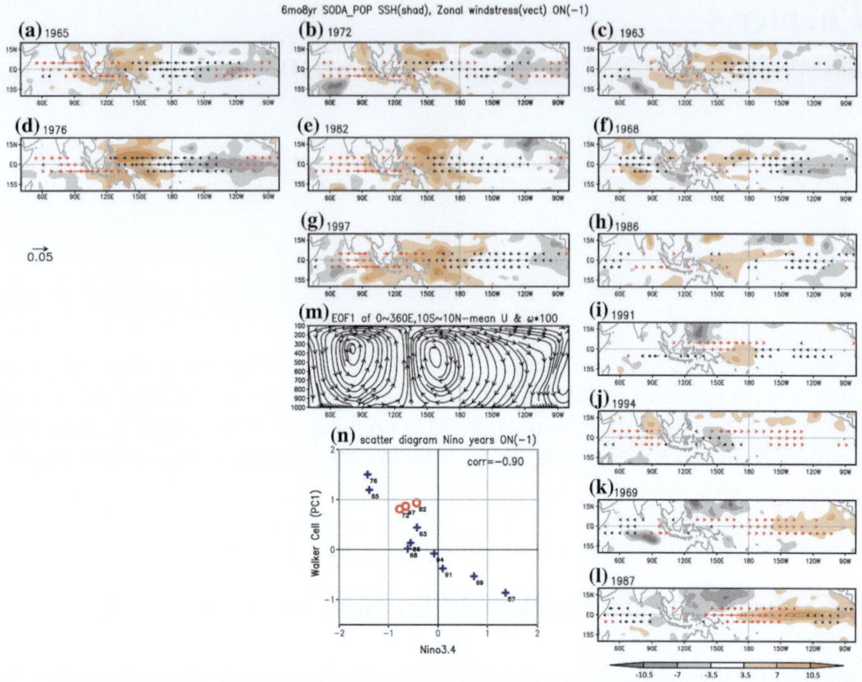

Fig. 4.1 **a–l** Anomalous SSH (*shading*; cm) and zonal wind stress (only wind speed >0.004 N/m^2 is shown as *vector*; *red* indicates westerly wind) for 12 El Niño events in ON(−1). **m** The first EOF mode of zonal and vertical pressure velocity (averaged over 10°S–10°N, vertical velocity is multiplied by 100 times) in the tropical belt during ON for 1959–2001; only the sector of 40°E–80°W is shown. **n** Scatter diagram of normalized Niño3.4 (*abscissa*) and normalized Walker cell index (the principle component of EOF1, *ordinate*) for 12 El Niño years in ON(−1); *Red circle* denotes super El Niño and *cross* denotes regular El Niño. The correlation is calculated using 12 El Niño events

content anomalies in the western Pacific. It is entirely possible that a few of them would be triggered by fast perturbations like typhoons (Keen 1982) or the MJO (Eisenman et al. 2005) for the sake of triggering the Bjerknes feedback. Comparing to the long timescale of ENSO, these types of triggers should be regarded as stochastic and thus they offer very limited predictability. The second column contains all known super El Niño events. Their negative heat content anomalies are less organized in the eastern tropical Pacific and more distant from the date line compared to the members in the first column. Also, the cold SST anomalies in the central tropical Pacific are relatively weak to the SST anomalies east of 120°W, although this pattern is not that significant in 1982 (Fig. 4.2). It would be tempted to label this kind of La Niña as the eastern Pacific (EP) La Niña, but, as Kug and Ham (2011) pointed out, types of La Niña are much difficult to be distinguished. Although the La Niña preceded super El Niño is weak, however, the convergence of the anomalous zonal wind stress in the western Pacific/Indian Ocean sector

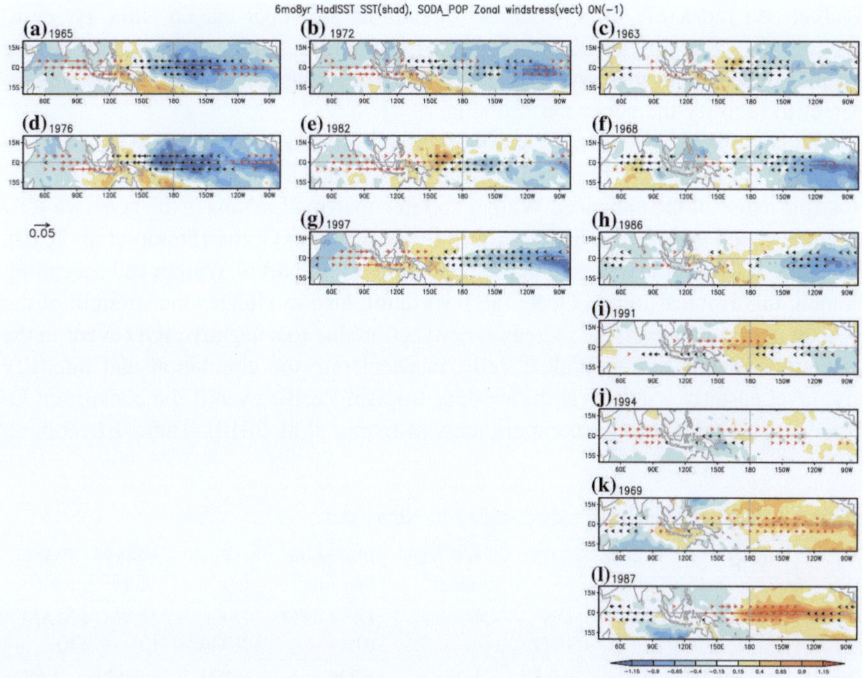

Fig. 4.2 **a–l** Anomalous SST (*shading*; °C) and zonal wind stress (only wind speed >0.004 N/m^2 is shown as *vector*; *red* indicates westerly wind) for 12 El Niño events in ON(−1)

displays a comparable accelerating Walker cell in comparison to their counterpart in 1965, 1976 El Niño events. Therefore, positive heat content anomalies are also largely accumulated in the western tropical Pacific by the tropical Pacific surface easterlies.

Since the super-ma La Niña is suggested as a precondition of super El Niño, is it possible to identify the super-ma La Nina? Note that it is the mildness of La Niña with a strong Walker cell characterizing super-ma La Niña. However, conventional measurements on ENSO strength such as Niño-3.4 SST index are closely correlated with the intensity of Walker cell due to the Bjerknes instability. Therefore, mild La Niña events as in 1982 and 1997 ON(−1) are unlikely to generate strong Walker cells, and not surprisingly these Niño indices appear less capable of isolating this feature. To address this issue, a Walker cell index is designed to identify super-ma La Niña based on the EOF analysis of zonal and vertical winds over the tropical cross section (averaged over 10°S–10°N along 0°–360°E). Figure 4.1m shows the first EOF mode that describes the Walker cell, accounting for 53.5 % of the total variance. In Fig. 4.1n the Walker cell index is plotted against Niño-3.4 index, both normalized, for all Niño (−1) years. Super El Niño events (red circles) congregate tightly within a narrow segment (0.81 standard deviations, 0.93 standard deviations) of the Walker cell index, which implies that only the La Niña events with the

Walker cell index of ~0.85 standard deviations are super-ma La Niña. By comparison, the Niño-3.4 index of super-ma La Niña does not fall in such a unique manner into a short segment. They are more scattering and thus relatively incompetent to identify the super-ma La Niña.

The slightly inconsistency of the Walker cell intensity and the ENSO strength in super-ma La Niña may be resulted from forcing outside the tropical Pacific. One possible cause of the enhanced Walker cell in super-ma La Niña is the negative IOD with increased convection in the eastern tropical Indian Ocean (Izumo et al. 2010). Since the IOD has the largest variations in the western part of Walker cell ascending branch, this tropical Indian Ocean SST variation also modulates the strength of the Walker cell. As a result, the increased convection due to a negative IOD event in the ascending branch of the Walker cell can accelerate the circulation and intensify low-level easterly winds over the western tropical Pacific even if the concurrent La Niña is weak (see numerical experiments in Izumo et al. 2010). Table 4.1 (column

Table 4.1 Various normalized indices for 12 El Niño years

	Niño-3.4[a]	Walker Cell[b]	IOD[c]	TWP heat content[d]	Philippine Sea low[e]	Hawaii low[f]	PMM SST[g]	PMM wnd[h]
	ON(−1)	ON(−1)	ON (−1)	ON(−1)	D(−1)JFM (0)	D(−1) JFM(0)	MAM (0)	MAM (0)
1972	**−0.78**	**0.81**	**−0.58**	**1.27**	**−0.82**	**−2.71**	**2.12**	**1.97**
1982	**−0.43**	**0.93**	**-0.70**	**1.72**	**−0.90**	**−1.18**	**2.03**	**1.71**
1997	**−0.66**	**0.87**	**−2.44**	**1.67**	**−1.66**	**−1.02**	**1.92**	**2.68**
1963	−0.43	0.44	0.00	0.67	−0.39	0.11	0.67	0.46
1965	−1.39	**1.19**	**−1.27**	0.74	**−1.01**	**−0.82**	−0.46	0.31
1968	−0.62	0.01	1.05	0.32	−0.37	**−1.86**	1.21	1.04
1976	−1.42	**1.50**	**−1.30**	**1.89**	**−1.11**	−0.54	0.11	0.81
1986	−0.56	0.13	0.02	0.26	**−0.86**	0.21	1.29	0.96
1991	0.09	−0.38	0.33	−0.23	0.21	−0.52	0.47	0.21
1994	−0.08	−0.08	-0.08	−0.35	−0.62	0.97	−0.24	−0.31
1969	0.72	−0.54	**−1.00**	−0.71	0.60	**−1.42**	0.79	0.33
1987	1.36	−0.86	0.25	−0.91	1.65	0.25	0.49	0.76

*Before calculate Niño-3.4, TWP Heat Content, Philippine Sea Low and Hawaii Low indices, monthly data is the first averaged over-defined region and then normalized, and these indices are calculated as bimonthly/4-month mean data. Walker cell and PMM indices are derived using yearly data and then normalized

[a]Niño-3.4: SST anomaly averaged over 5°S–5°N, 120°W–170°W

[b]Walker cell: principle component of EOF 1 of 1000–100 hPa zonal and vertical velocity in the tropical belt during ON

[c]IOD: difference of SST anomaly between 10°S–10°N, 50°E–70°E, and 0°S–20°S, 90°E–110°E

[d]TWP Heat Content: SSH anomaly averaged over 130°E–180°, 5°N–15°N, and 5°S–15°S

[e]Philippine Sea Low: SLP anomaly averaged over 5°N–15°N, 130°E–150°E

[f]Hawaii Low: 925 hPa stream function anomaly averaged over 10°N–25°N, 170°E–160°W

[g]PMM_SST: SST expansion coefficient of SVD2 over 20°S–20°N, 150°E–120°W in MAM(0)

[h]PMM_wnd: wind stress expansion coefficient of SVD2 over 20°S–20°N, 150°E–120°W in MAM (0)

3) shows that the normalized IOD index in ON(−1) of three super El Niño events are less than −0.5 standard deviations and indicates that a super-ma La Niña coexists with a negative IOD event. In contrast, the IOD index in other four regular El Niño with a mild La Niña in ON(−1) (1963, 1968, 1986, and 1994) are either near zero or even positive, so the Walker cell is only slightly enhanced and these La Niña events cannot be recognized as super-ma La Niña.

Nevertheless a super-ma La Niña only provides a necessary condition preceding a super El Niño event, since Fig. 4.1n just covers those La Niña years followed by an El Niño in the winter of next year. If all La Niña years are included, twice as many La Niña years will occupy the same super-ma La Niña segment. Therefore, by screening the Walker cell index alone cannot guarantee the birth of super El Niño. It takes another precondition in the form of stationary perturbations from middle latitudes in the North Pacific and from the western North Pacific to pin down the coming of a super El Niño, as shown in the next section.

4.2 Precondition 2: Simultaneous Emergence of the Hawaii Low and Philippine Sea Low

The coexistence of the Hawaii low and Philippine Sea low in the previous winter of El Niño is suggested to be the second precondition of super El Niño. Figure 4.3 shows anomalous SLP fields during D(−1)JFM(0) for 12 El Niño events. The most striking and unique feature of super El Niño (the second column) is that vast low SLP anomalies occupy the northern tropical Pacific. The huge low SLP anomalies comprise two low pressure systems, one centered over the Philippine Sea and another near the Hawaii. Here they are referred to as "Philippine Sea low" and "Hawaii low," respectively. The Philippine Sea low is reminiscent of the opposite phase of the Philippine Sea anticyclone (Wang et al. 2000), which may be induced by a La Niña via in situ SST over the western North Pacific (Wang and Zhang 2002), tropospheric temperature (Hsieh 2010, Master's thesis), or remote forcing from the tropical Indian Ocean SST, which is basically passively controlled by the eastern equatorial Pacific SST (e.g., Watanabe and Jin 2002; Watanabe 2003; Wu et al. 2009). Therefore, the Philippine Sea low also can be observed in the winter before 1965, 1976, and 1986. The Hawaii low bears a resemblance of the southern lobe of the NPO. However, in contrast to a typical NPO as shown in 1963, 1965, and 1968 El Niño events, the Hawaii low in all super El Niño events is characterized by a low-latitude position that is much south than the southern lobe of the NPO. The center of the Hawaii low is located southwest to the Hawaii islands; the northern edge is below 30°N; besides, its circulation extends southwestward to the tropics around the date line. Although the Hawaii low and Philippine Sea low are combined and located south of a North Pacific SLP dipole in the winter before the 1997 super El Niño, this vast low pressure is also in between 30°N and the equator.

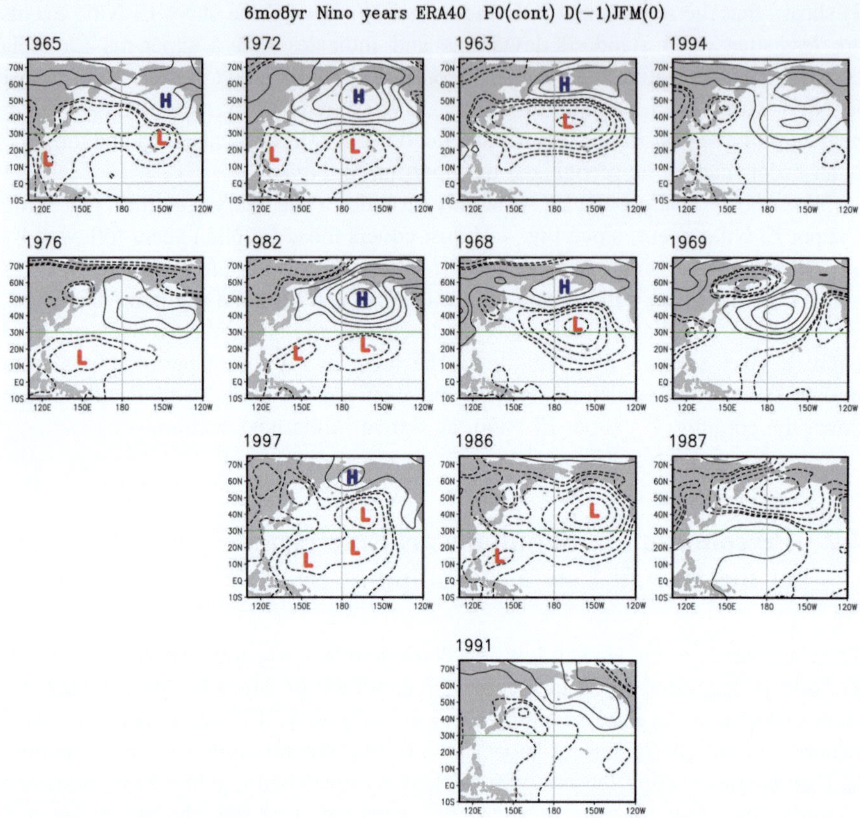

Fig. 4.3 Anomalous SLP (contour levels are −5, −3.5, −2, −1.2, −0.6, −0.3, 1, 2, 3.5, 5, and 7 hPa) for 12 El Niño events in D(−1)JFM(0). The letter *H* (*L*) denotes the center of high (low) SLP anomalies

In early winter under the super-ma La Niña condition, the Hawaii low and Philippine Sea low establish separately (or autumn for 1997 super El Niño event, not shown). However, two systems gradually develop into the PMM and form a key box with enhanced deep convection in the north of central equatorial Pacific that facilitates SHB when super El Niño onset. To illustrate how the Hawaii low and Philippine Sea low evolve into the PMM and the underlying mechanism led to the key box, two low pressure systems are separately discussed and their net contributions are presented in the following two subsections.

4.2.1 Sprawling of the Hawaii Low

The Hawaii low is relatively isotropic when it first forms (black contours in Figs. 4.4c, g and i), and it exhibits an equivalent barotropic structure (Fig. 4.5). The vertical structure and its location, which is to the north of the critical line (where zonal wind = 0, a necessary condition for stationary Rossby waves to propagate, James 1995), imply that it may be a result of stationary Rossby wave trains originated from high latitudes. Instead of keeping isotropic, the Hawaii low continues intensifying and expanding, and becomes more spatial asymmetric. As depicted in the schematic diagrams from Fig. 4.4a–b, it deforms and elongates along the NE–SW axis, so the southwest portion of its cyclonic circulation also extends southwestward toward the equator. This evolution can also be seen in super El Niño composite and individual events (Fig. 4.4c–j). The evolution of the Hawaii low resembles a galaxy "sprawling" and stretching its "leg," southwesterly winds, toward the central equatorial Pacific. Consequently, the sprawling of the Hawaii low is a crucial step for this external forcing from subtropics to directly influence the equatorial Pacific (ENSO source region) and trigger El Niño.

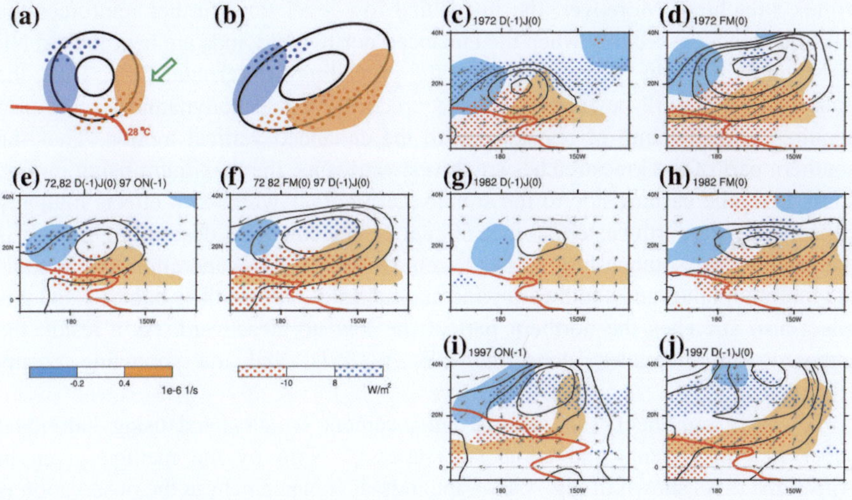

Fig. 4.4 **a–b** Schematic diagram illustrating that Hawaii low sprawls and its structure tilts in NE–SW axis via the mutual enhancement of the Sverdrup balance and the WES feedback on vertical motion. *Black contours* denote low SLP anomalies. *Shading* denotes vertical motion induced by Sverdrup balance (ascending motion is shown by *orange*, and descending is shown by *blue*); *dots* denote anomalous latent heat flux (W m^{-2}) and positive value indicates that surface is heated. *Red line* indicates climatological SST 28 °C isotherm. In (**a**) *green vector* denotes climatological northwesterly winds. **e–f** are the same as (**a**)–(**b**) but for three super El Niño events composites. *Vectors* indicate anomalous wind stress (wind stress value exceeds 0.01 N m^{-2} is shown); the *contour* levels for SLP are −5, −3.5, −2, and −1.2 hPa. **c**, **d**, **g–j** Same as (**e**)–(**f**) but for three super El Niño events

For a given stationary pattern coming out in the northern winter, two effects are held most responsible for breaking down the spatial symmetry. One is the WES feedback whereby the NPO can migrate to the tropics as proposed in the SFM (Vimont et al. 2009), and the other is the Sverdrup balance. The Sverdrup balance states that off equator the vortex stretching term is balanced with the horizontal advection of earth vorticity (Hoskins and Karoly 1981). Thus, in the Northern Hemisphere ascending motion will be found on the right side and descending motion appears in the left side over a surface low anomaly (e.g., Nigam and Chan 2009). Here we just take down that both Sverdrup and WES feedback could reshape vertical motion by dynamic forcing (vortex stretching) and thermodynamic forcing (convection/SST linkage).

Suppose initially a low-level anomalous low is imposed on an area of the North Hemisphere outside the equator. Assume it starts with a relatively isotropic shape. The interesting part occurs when these two effects are overlapped around the southeastern and northwestern corners, as sketched in Fig. 4.4a. Taking the southeastern corner as example, the vortex stretching (orange shading) will gain some extra booster, i.e., extra ascending motions, if convection becomes active over the warm SST which has been induced by reduction of scalar wind (orange dots). Hence, a new round of Sverdrup balance will be required. This time the causality will be seen as additional northward winds near surface are brought by this extra vortex stretching. Moreover, the intensified low-level flow further reinforces the WES feedback, especially when the enhanced northward winds are bent toward NE direction by Coriolis torque, opposing the prevailing background flow. Thus, the underlying SST will be further warmed up. In turn, thermodynamic forcing also acquires a fresh round of booster due to the enhanced vertical motion. Thus the southern part of the low stretches southwestward since the maximum rising motion shifts from the eastern side to the southeastern corner where two effects mutually enhance. In the northwestern corner mutual enhancement on descending motion is the same as the southeastern corner except for increased evaporation which cools the SST anomalies and southward wind anomalies in the Sverdrup balance, and this effect also stretches the northern part of the low northeastward. As a result, the imposed low elongates along the NE–SW axis, and the sprawling occurs (Fig. 4.4b).

To verify our theory, the mutual enhancement is calculated using individual super El Niño events and composite super El Niño by the method given in Appendix 1. As shown in Fig. 4.4c–j, the mutual enhancement in the observation is generally in agreement with our theory, although the shape is more elliptic and the overlapped region shifts to the southern side due to the southerly winds over the north eastern Pacific induced by super-ma La Niña. The position of the Hawaii low facilitates the mutual enhancement works more efficiently in the winter before super El Niño events than other El Niño events because it occurs over the region where the efficiency of SST to activate convection is high. As shown in Fig. 4.4e, the average latitude of the Hawaii low southern edge is around 10°N. Since the SST in this area lingers below the threshold of deep convection limit (28 °C climatological SST isotherm as proxy), the SST is easily raised up to deep convection threshold

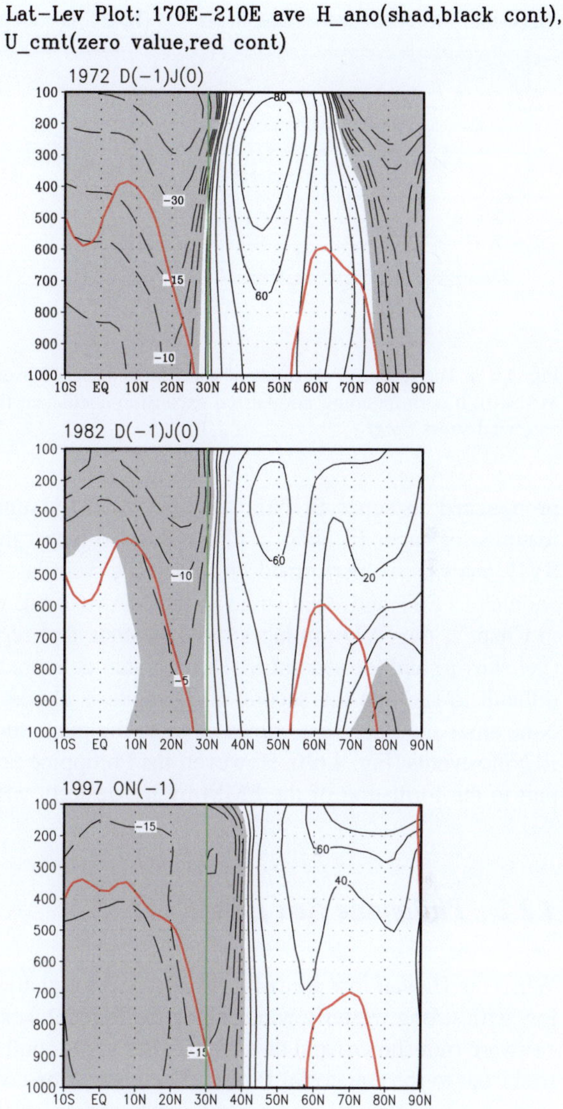

Fig. 4.5 Latitude-height cross section of anomalous geopotential height (*contour*, m; negative value is *shaded*) and zero value of climatological zonal winds (*red contour*) averaged over 170°E–150°W for three super El Niño events

via the WES feedback. In contrast, for a NPO-generated low pressure, the southern edge is relatively far from the central tropical Pacific and the deep convection threshold (Fig. 4.3), so the efficiency of SST anomaly to active convection is low. Besides, it takes more time for a NPO-generated low pressure anomaly to move to the tropics where oceanic Kelvin waves can be driven to kick off an El Niño event.

A significant result of the Hawaii low sprawling is marked meridional SST and wind anomaly bands structure across the central-eastern tropical Pacific, i.e., a PMM, in the spring as shown in Fig. 4.6a. We find that the PMM is extremely

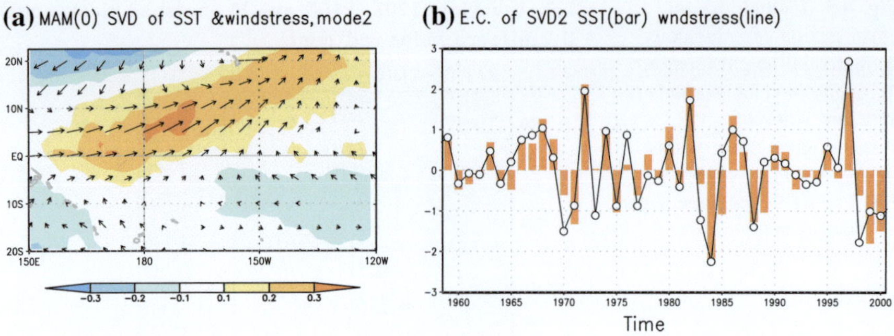

Fig. 4.6 **a** The second SVD mode of wind stress and SST over the central tropical Pacific during MAM(0); **b** corresponding normalized expansion coefficient time series of SVD2 for SST (*bar*) and wind stress (*line*)

pronounced in super El Niño events, which is suggested as one of the unique features for super El Niño in Chap. 3.2. To recast the PMM in our framework, a SVD analysis is performed in the same vein as Chang et al. (2007). Using anomalous SST and wind stress in the MAM(0) but with filtered data as described in Chap. 2, the SVD produces the first mode that represents the post-Niño pattern (not shown) and the second SVD mode that delivers the PMM (Fig. 4.6a). It is not difficult to see that the second SVD mode is almost tailor-made for the super El Niño since no other year comes close to the magnitude attained by the three super El Niño events (Fig. 4.6b). However, the Philippine Sea low also plays an important part to the formation of the PMM by its westerly winds.

4.2.2 Philippine Sea Low

When the Hawaii low sprawls and affects the equator in FM(0), the Philippine Sea low with strong westerly winds along the tropical western North Pacific also moves eastward (not shown) and thus contributes to the southwest portion of the PMM that sticks out to the equatorial Pacific. The circulations of two low systems merge and become a massive low across the northern tropical Pacific; thus the merged low produces a strong low-level westerly wind belt along the north flank of the western and central equatorial Pacific. The oceanic mixed layer heat budget analysis, introduced in Appendix 2, indicates that the westerly wind belt drives two dominant mechanisms to the warming tendency over the southwest portion of the PMM, which is over the north-central equatorial Pacific (0°–7.5°N, 155°E–170°W), during D(−1)JFM(0). As shown in Fig. 4.7, the major contribution is the surface latent heat flux associated with the wind speed variation (Fig. 4.7c), which warms the temperature about 0.27 °C per month (the first term on the RHS), and the second contribution is the zonal advection of mean temperature by anomalous zonal

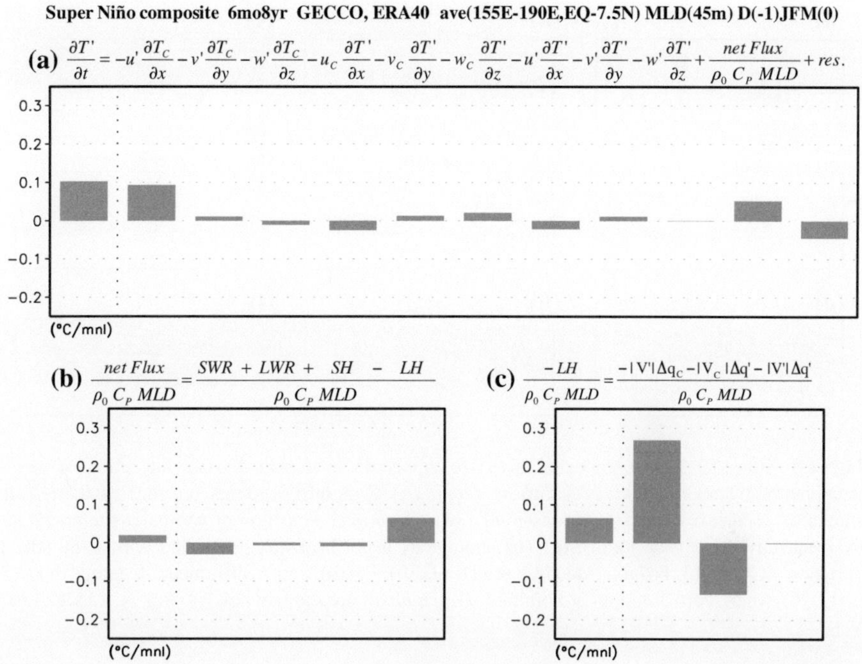

Fig. 4.7 **a** Individual terms in the mixed layer heat budget of super El Niño composite for north-central equatorial Pacific (0°–7.5°N, 155°E–170°W) during D(−1)JFM(0) using mixed layer depth of 45 m (*unit*°C/month). **b** Contribution from anomalous surface heat flux (*left-hand side*), which decomposed as the anomalous surface short wave radiation, long wave radiation, sensible heat flux and latent heat flux (*right-hand side)* using ERA-40 data; **c** Contribution of anomalous surface latent heat flux (*left-hand side*), which linearly decomposed into components that related with wind speed variation, moisture difference variation, and their nonlinear contribution

currents; this term contributes about 0.1 °C per month (Fig. 4.7a, the first term on the RHS). The results are similar no matter the calculation is using super El Niño composite or individual super El Niño cases. The largest warming effect is consistent with the SFM. The westerly winds of the merged low system decrease the wind speed, and thus the reduction of upward latent heat flux warms the mixed layer temperature, and this wind belt also partly drives the zonal advection. Figure 4.8a–b indicates that strong, organized but asymmetric (to the equator) anomalous mixed layer oceanic eastward currents establish in D(−1)J(0) and prevail in the whole northern central and western tropical Pacific during FM(0). Accordingly, the anomalous zonal currents transport climatological mean high temperature water from the western Pacific to the date line and warm this region (green box). The strong eastward currents are even more pronounced in AM as shown in composite (Fig. 4.8c) and each super El Niño case (Fig. 3.9). However, the zonal advection is mainly effective between 5°N and the equator, since the zonal gradient of mixed layer climatological temperature only reside in this sector (not

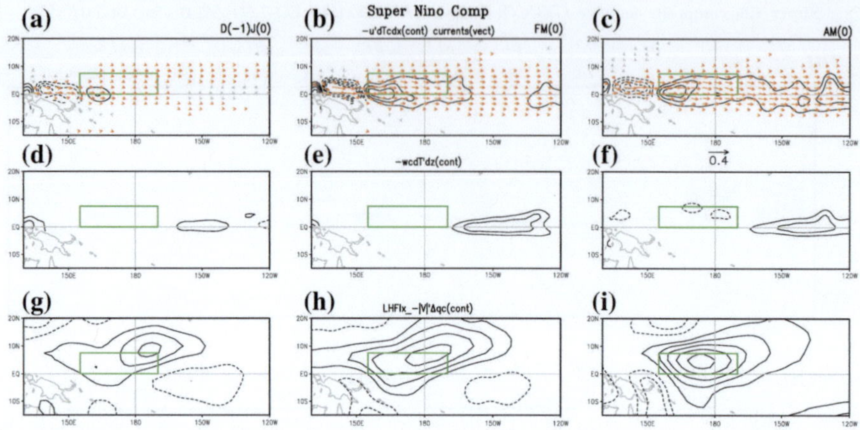

Fig. 4.8 Super El Niño composite of (**a**)–(**c**) upper 45 m averaged zonal advection of mean temperature by anomalous zonal current (*contour*; °C/month), currents (current >0.02 m/s is shown by *vector*; *orange* denotes eastward); (**d**)–(**f**) vertical advection of anomalous temperature by mean upwelling (°C/month); (**g**)–(**h**) anomalous latent heat flux associated with wind speed variation (°C/month). From *left* to *right* are D(−1)J(0), FM(0), and AM(0) mean. Contour interval is 0.1 °C/month; zero contours are omitted. *Green boxes* are the key box (0°–7.5°N, 155°E–170° W)

shown). Note that the eastward currents consist not only of Ekman flows, but also of geostrophic flows. Although the geostrophic flows are not directly produced by the westerly winds of the Hawaii low and Philippine Sea low, these flows are associated with the super-ma La Niña via the positive upper-ocean heat content anomalies, which are first induced in the western tropical Pacific and then gradually propagate into the equator in D(−1)JFM(0) (not shown), forming the eastward geostrophic flows. As a result, the coexistence of the Philippine Sea low and the Hawaii low contributes to the extension of the PMM to the equatorial Pacific in terms of intensified westerly winds and warm SST anomaly. Furthermore, the SST anomalies in green box raised to a prominently positive value during AM(0) as shown in Fig. 4.9. (The development of the 1982 super El Niño event slightly lags the other two super El Niño events by 1–2 months. Thus the SST anomaly in 1982 is shown using May–June mean.) Some regular El Niño events such as the 1969 and 1987 El Niño also have high SST anomaly in the green box; however, the warm SST anomaly results from the previous El Niño condition; the 1991 El Niño event has this signature as well, but there is no corresponding PMM to sustain the pattern. The linkage between the PMM and the triggering of ENSO is well established by many studies (e.g., Chang et al. 2007; Zhang et al. 2009); however, we want to point out that the warm SST anomaly in the green box is a vital step for an El Niño to grow into a super El Niño, which will be elaborated in Sect. 4.3 when the onset is introduced.

One question that may be raised is: Since the Hawaii low and Philippine Sea low are important to the PMM and the triggering of super El Niño, does this

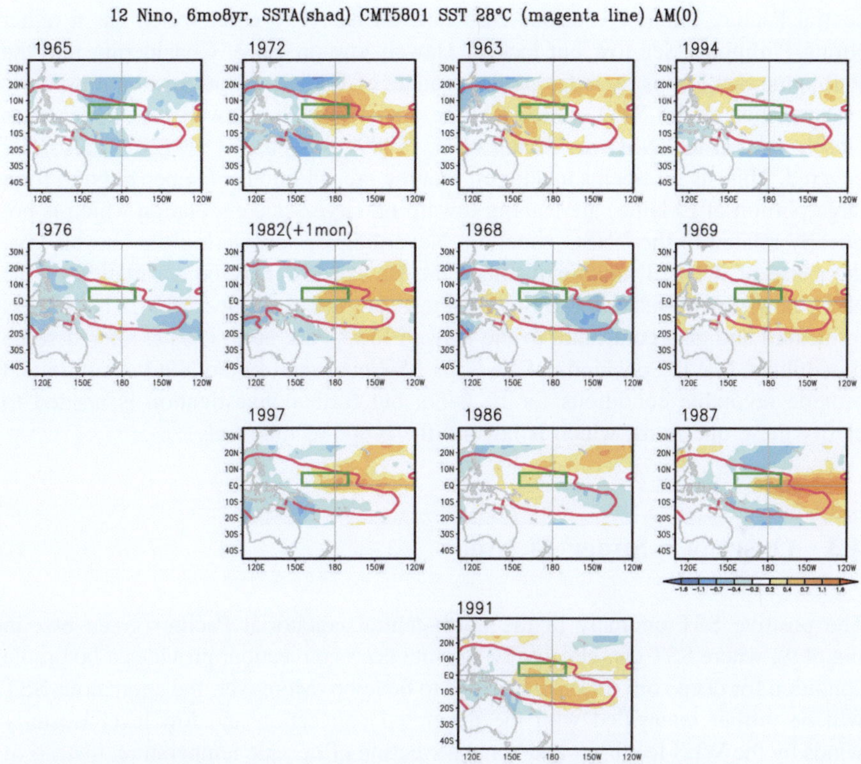

Fig. 4.9 Anomalous SST (*shading*, °C) and climatological SST 28 °C isotherm (*magenta line*) for 12 El Niño events in AM(0), except for the El Niño 1982 in MJ(0)

simultaneous emergence take place very often? We use indices to examine this co-occurrence. The Philippine Sea low index is defined as time series of the SLP anomaly averaged over 5°N–15°N, 130°E–150°E during D(−1)JFM(0), and the Hawaii low index is defined as time series of 925 hPa stream function anomaly averaged over 10°N–25°N, 170°E–160°W to characterize its location and cyclonic circulation in the northern tropical Pacific. Normalized indices for 12 El Niño events are listed in Table 4.1 (the 5th and 6th columns). It shows that among nine regular El Niño cases during 1960–2001, only the 1965 El Niño has both Philippine Sea low and Hawaii low recorded comparable values to super El Niño events in D (−1)JFM(0). Nevertheless, the latitude of the southern pole of the NPO in 1965 D (−1)JFM(0) is relatively far from the central equatorial Pacific and its circulation is confined within northeast Pacific mid-latitudes. Hence, the anomalous westerly winds over northern tropical Pacific are absent (not shown) and the amplitudes of the PMM are small in spring (Fig. 4.6b). Furthermore, the sprawling process will no doubt be botched as in the case of 1965 El Niño when the initial SST in the central equatorial Pacific was found too cold to kick off convection. Only a super-ma La Niña can be susceptible to sprawling due to the mild or neutral SST condition in the central Pacific. The very curious case falls on the 1976 case, which has a rather strong Philippine Sea low but lack of Hawaii low presence. Considering that the Philippine Sea low is regarded as the product of ENSO, extratropical atmospheric systems such as the NPO, which may be a source of the Hawaii low, appear less closely related to ENSO, as Zhang et al. (2009) and Park et al. (2012) also discovered. This, again, seems to suggest that the preconditions of super El Niño (or a large portion of El Niño) are leaning toward on large-scale circulation which is not linearly related to the ENSO situation. Nevertheless, considering the Hawaii low, the cause of it is still unknown. Since the Hawaii low may be a variation of the NPO, how can we relate these two systems? Beside, does super-ma La Niña exert influences and also contribute to the Hawaii low? Park et al. (2012) offered some possibilities that the positions of La Niña SST may modify the NPO structure and provide favorable conditions for El Niño, but further investigation is needed to clarify these questions, which is beyond the scope of this study.

4.3 Onset of a Super El Niño

The positive SST anomaly in the north-central equatorial Pacific (green box in Fig. 4.9), where SST can effectively activate deep convection, provides a favorable condition for deep convection anomalies to develop. Moreover, the anomalous SST will be further intensified with the generated convection and low-level westerly winds by the WES feedback and zonal advection of oceanic temperature, instead of returning to a neutral state. Hence, a sudden explosion of enhanced deep convection anomalies (using 200 hPa divergence anomalies as proxy) occurs in the this box in the mature stage of the PMM, AM(0), and produces anomalous high-level divergent flows toward south Australia, thus leading to an anomalous local Hadley cell between the ENSO source region and the Southern Hemisphere as shown in Fig. 4.10. The enhanced convection in this box is crucial for the onset of super El Niño for two reasons. First, the associated anomalous westerly winds of the enhanced convection over this region can trigger the Bjerknes feedback, the primary growing process of ENSO. Second, the induced local Hadley cell is the "embryo" of the SHB that will serve as a booster to accelerate El Niño's growth in the following season, leading to a super El Niño. Therefore, the box is a key region for super El Niño. The formation of the local Hadley cell may be related to the seasonal transition of the climatological local Hadley cell. Since late spring is the time for the Southern Hemisphere to proceed to winter, and the enhancement of the central equatorial convection may intensify the climatological local Hadley cell and thus subsides over the climatological local Hadley descending branch. Once the local Hadley cell forms, since the Southern Hemisphere subtropical jet begins to accelerate and the absolute vorticity meridional gradient becomes intense from AM, the high-level divergent winds induced by the enhanced convection can effectively generate Rossby wave source and establish a surface high anomaly via the Sverdrup

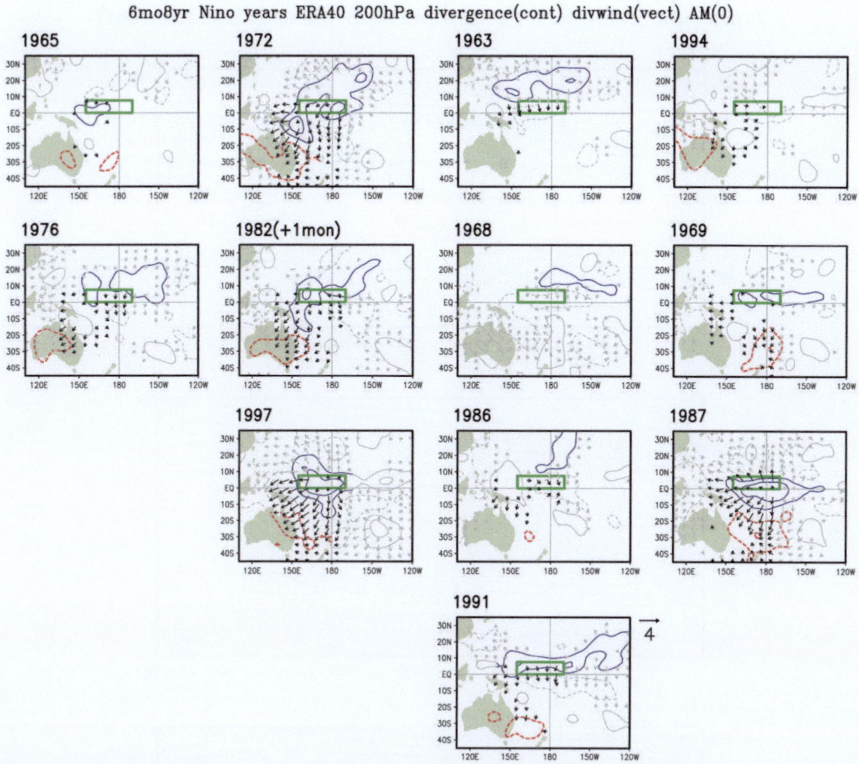

Fig. 4.10 Same as Fig. 4.9 but for anomalous 200 hPa divergence (*contour* levels are −2, −0.75, 0.75, 2; unit:1e-6 s^{-1}), and 200 hPa divergent winds (*vector*, m/s, only magnitude of wind exceeds 0.6 m/s is drawn). Anomalous divergence in the central Pacific, convergence near Australia, and divergent winds in between are plotted in *colored contours* and *black arrows*

balance to strengthen the local Hadley cell (Fig. 4.11). Therefore, it quickly develops into the SHB. More details about the SHB will be shown in Chap. 5.

The onset of super El Niño is not only characterized by the key box and the SHB. As the vast low pressure system in the northern tropical Pacific warms the key box, the eastern equatorial Pacific, where Niño-3 SST index is defined, warms up at the same time. Figure 4.8e–f indicates that the mixed layer ocean temperature is warmed due to the vertical advection of anomalous temperature by climatological upwelling. The anomalous warm water in the eastern equatorial Pacific subsurface comes from the western tropical Pacific where a large amount of warm water was accumulated in the previous year due to super-ma La Niña. The warm water mass first converged into the equator via coastal Kelvin waves (Philander 1990), and once reached the equator, it turned 90° and was transported by the rapid undercurrents along the thermocline to eastern equatorial Pacific by FM(0) (Fig. 3.12). Since the thermocline is shallow and near the mixed layer in the eastern equatorial Pacific, the subsurface warm water is brought up via the climatological upwelling to

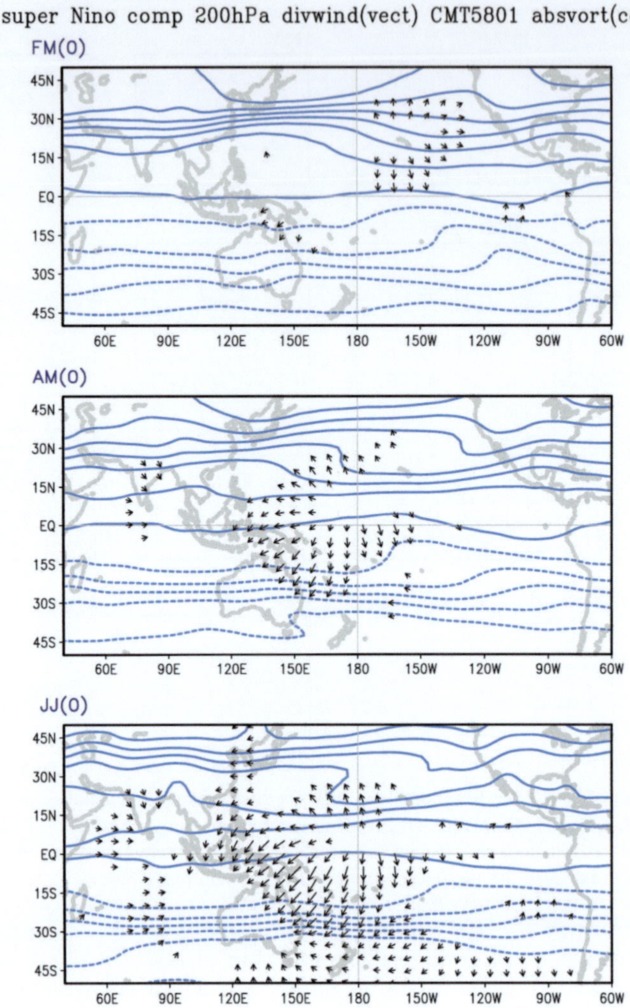

Fig. 4.11 Climatological absolute vorticity (*contour* with interval = 2e5 s^{-1}) and anomalous 200 hPa divergent winds of super El Niño composite (only wind speed >0.2 m/s is shown by *vector*)

the mixed layer and this effect also contributes to the onset of super El Niño in AM (0).

Consequently, the onset of super El Niño is triggered via two paths under the surface and summarized as shown in Fig. 4.12. In early winter, the Philippine Sea low, a response of La Niña, and the Hawaii low, an external forcing from North Pacific, simultaneously establish (Fig. 4.12a). The Hawaii low gradually sprawls via the Sverdrup balance and the WES feedback; the Philippine Sea low also moves eastward, so two low systems become a massive low with a strong low-level

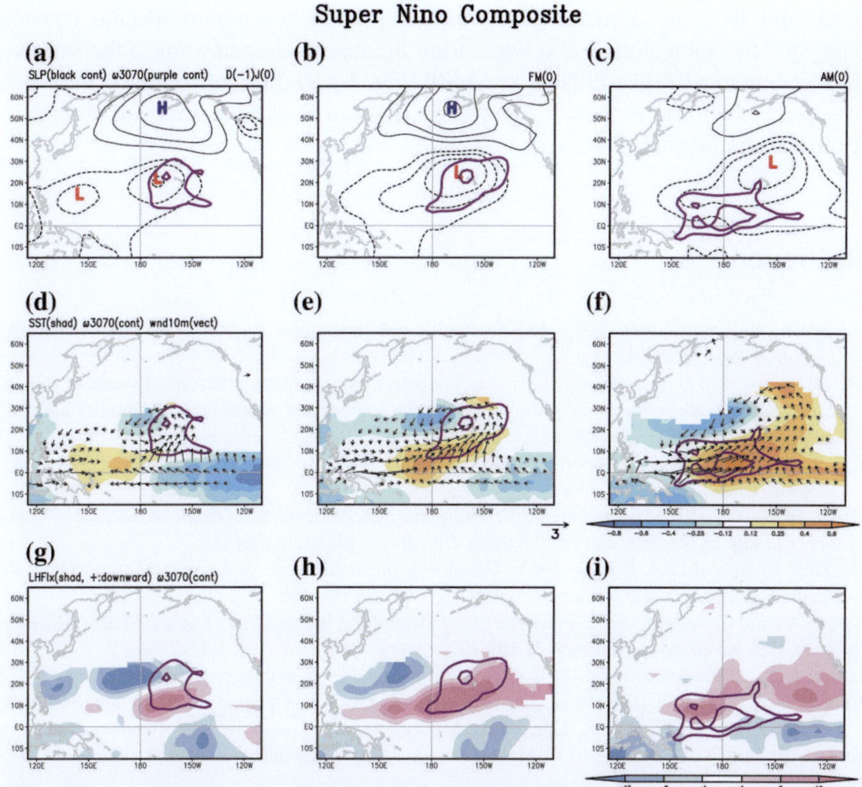

Fig. 4.12 Super El Niño composite of anomalous (**a**)–(**c**) SLP (*black contour*; levels are −1.2, −0.8, −0.4, 1, 2, and 3.5 hPa); (**d**)–(**f**) SST (*shading*; °C) and 10-m wind fields (*vector*; m/s); (**g**)–(**i**) surface latent heat flux (*shading*, W/m^2, positive values denote downward). The shading and vectors are only shown in the region of low SLP anomaly and tropics in (**d**)–(**i**). *Purple contours* are anomalous vertical pressure velocity averaged over 300–700 hPa (contour levels are −0.012 and −0.024 Pa/s); only ascending motion associated with Hawaii Low is shown

westerly wind belt along the north of western and central equatorial Pacific (the first two columns in Fig. 4.12). Furthermore, because of the nearly neutral central SST anomalies in the super-ma La Niña (Fig. 4.12d), the WES feedback and zonal advection by anomalous currents (not shown) easily turn the SST anomalies in the central equatorial Pacific to positive. As a result, a box in the north of the central equatorial Pacific sets up a perfect condition to initiate the SHB, and a Central Pacific (CP)-like El Niño occurs early in FM(0). This path is a bit like the advective-reflective oscillator, which emphasizes the importance of warm pool eastward extension by oceanic waves to the onset of ENSO. However, in this study, it is suggested that the enhanced convection and warm SST anomalies "within" the climatological warm pool region induce the Bjerknes feedback and the SHB thus contributing to the onset of super El Niño. Another path follows the recharge oscillator theory whereby the equatorial Pacific is recharged due to the super-ma La

Niña, and then the climatological upwelling along the eastern Pacific equator transports the anomalous warm water from the thermocline upward to the surface, thus leading to a EP-like El Niño in AM(0) (Fig. 4.12f). In a consequence, the onset of super El Niño is composed of a strong CP and EP El Niño structure, a distinct feature in the beginning stage.

References

Chang, P., and Coauthors, 2007: Pacific meridional mode and El Nino-southern oscillation. *Geophys. Res. Lett.*, **34.**

Chiodi, A. M., and D. E. Harrison, 2007: Mechanisms of summertime subtropical southern Indian ocean sea surface temperature variability: On the importance of humidity anomalies and the meridional advection of water vapor. *J. Clim.*, **20,** 4835-4852.

Eisenman, I., L. Yu, and E. Tziperman, 2005: Westerly Wind Bursts: ENSO's Tail Rather than the Dog? *J. Clim.*, **18,** 5224-5238.

Hsieh, Pei-yun (2010) Maintenance of the Philippine Sea Anticyclone *(Graduate institute of Atm. Sci. College of Science, National Taiwan University,* Master's thesis).

Hoskins, B. J., and D. J. Karoly, 1981: The steady linear response of a spherical atmosphere to thermal and orographic forcing. *J. Atmos. Sci.*, **38,** 1179-1196.

Izumo, T., and Coauthors, 2010: Influence of the state of the Indian Ocean Dipole on the following year/'s El Nino. *Nature Geosci*, **3,** 168-172.

James, I. N., 1995: *Introduction to Circulating Atmospheres.* Cambridge University Press.

Keen, R. A., 1982: The Role of Cross-equatorial Tropical Cyclone Pairs in the Southern Oscillation. *Mon. Wea. Rev.*, **110,** 1405-1416.

Kug, J.-S., and Y.-G. Ham, 2011: Are there two types of La Nina? *Geophys. Res. Lett.*, **38,** L16704.

Li, T., Y. S. Zhang, E. Lu, and D. L. Wang, 2002: Relative role of dynamic and thermodynamic processes in the development of the Indian Ocean dipole: An OGCM diagnosis. *Geophys. Res. Lett.*, **29.**

Nigam, S., and S. C. Chan, 2009: On the Summertime Strengthening of the Northern Hemisphere Pacific Sea Level Pressure Anticyclone. *J. Clim.*, **22,** 1174-1192.

Park, J.-Y., S.-W. Yeh, J.-S. Kug, and J. Yoon, 2012: Favorable connections between seasonal footprinting mechanism and El Niño. *Clim Dyn*, 1-13.

Philander, S. G. H., 1990: El Niño, La Niña, and the Southern Oscillation.International Geophysics Series, Vol. 46, Academic Press, 293 pp.

Vimont, D. J., M. Alexander, and A. Fontaine, 2009: Midlatitude Excitation of Tropical Variability in the Pacific: The Role of Thermodynamic Coupling and Seasonality*. *J. Clim.*, **22,** 518-534.

Wang, B., and Q. Zhang, 2002: Pacific-east Asian teleconnection. Part II: How the Philippine Sea anomalous anticyclone is established during El Nino development. *J. Clim.*, **15,** 3252-3265.

Wang, B., R. Wu, and X. Fu, 2000: Pacific–East Asian Teleconnection: How Does ENSO Affect East Asian Climate? *J. Clim.*, **13,** 1517-1536.

Watanabe, M., and F. F. Jin, 2002: Role of Indian ocean warming in the development of Philippine sea anticyclone during ENSO. *Geophys. Res. Lett.*, **29.**

Watanabe, M., 2003: A moist linear baroclinic model: Coupled dynamical-convective response to El Nino. *J. Clim.*, **16,** 1121-1139.

Wu, B., T. J. Zhou, and T. Li, 2009: Seasonally Evolving Dominant Interannual Variability Modes of East Asian Climate. *J. Clim.*, **22,** 2992-3005.

Zhang, L., P. Chang, and L. Ji, 2009: Linking the Pacific Meridional Mode to ENSO: Coupled Model Analysis. *J. Clim.*, **22,** 3488-3505.

Chapter 5
Southern Hemisphere Booster

The main finding in Hong et al. (2014) is that the extraordinary magnitude of super El Niño is facilitated by a Southern Hemisphere transverse cell, which acts as a booster to further accelerate the low-level westerly winds in the tropical western Pacific, thus resulting in a super El Niño. In this chapter, this boosting mechanism is presented.

During the El Niño developing stage (June–November, or JJASON), unique features can be identified outside the tropics in the super El Niño composite compared with the regular El Niño composite (Fig. 5.1c–d). Of particular interest are the intensified 10 m westerlies over the western tropical Pacific that are closely associated with a large-scale coherent pattern in sea level pressure (SLP), with high pressure over southern Australia and low pressure over the southern Pacific. These two systems form an SLP dipole in the southern mid latitudes, far from the low latitudes where the typical Southern Oscillation is located. The anomalous high pressure over Australia (hereafter Australian high) spins off strong equatorward flow along the east coast of Australia, which moves across the equator near New Guinea, and veers off to the east from a potential vorticity perspective (e.g., Rodwell and Hoskins 1995) to accelerate westerly winds in the central Pacific. The veering off is also facilitated by a NW–SE aligned tropical trough anomaly near Melanesia, a result mainly caused by Rossby response to El Niño's anomalous convection, in addition to some regional effects near the Solomon Sea (see numerical experiments in Annamalai et al. 2010). It is proposed that this extratropical forcing from the Southern Hemisphere adds on top of the Bjerknes instability. By strengthening the westerly wind in the ENSO source region, the anomalous circulation in the Southern Hemisphere acts as a booster to amplify El Niño's growth, a critical step toward becoming a super El Niño.

Stephens et al. (2007) pointed out that a standing wave pattern in SLP in the Southern Hemisphere, similar to the one in Fig. 5.1c, during strong El Niño developing stage weakens the annual cycle and drives the southwesterly wind toward the equatorial Pacific. They proposed that the South Pacific low-pressure anomaly plays an important role in regulating the strength of trade winds, which leads to changes in El Niño amplitude. In this study, however, we find that in the case of super El Niño events the Southern Hemisphere SLP dipole interacts with the

L.-C. Hong, *Super El Niño*, Springer Theses,
DOI 10.1007/978-981-10-0527-5_5

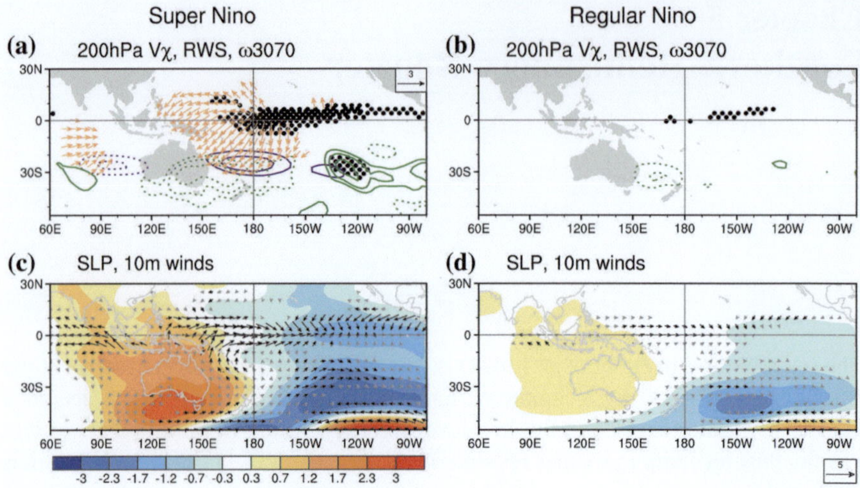

Fig. 5.1 JJASON(0) mean composite maps of (**a**) anomalous 200 hPa divergent winds (*vector*, only wind speed >1.1 m/s is shown), 200 hPa RWS in the Southern Hemisphere (*contour* interval: 1e-11 $1/s^2$; *green* denotes RWS induced by vortex stretching and *purple* denotes RWS caused by advection of vorticity via anomalous divergent winds; zero contours are omitted), vertical pressure velocity averaged over 300–700 hPa (*dots*; only upward motion <−0.012 Pa/s is shown) and (**c**) anomalous SLP (*shading*, hPa), 10-m winds (*vector*, only wind speed >0.4 m/s is shown; wind speed >0.8 m/s is plotted in *black*) for super El Niño composite. (**b**) and (**d**) are the same as (**a**) and (**c**), respectively, except for regular El Niño composite (Reprinted from Hong et al. 2014 with permission from John Wiley & Sons, Inc., New York, USA)

tropics in a more intimate way through a self-sustaining transverse cell that connects the El Niño source region and the Australian high.

To depict the transverse cell, anomalous upper level atmospheric fields and vertical motion in the upper-middle troposphere are shown in Fig. 5.1a. During super El Niño years, anomalous divergent winds generated from enhanced deep convection over the central equatorial Pacific move southwestward and then converge over the Maritime Continent and the Australian high (Fig. 5.2). The upper level atmosphere features a unique pattern of anomalous Rossby wave source (RWS) (see Sardeshmukh and Hoskins 1988 for the derivation of anomalous RWS). Anomalous divergent winds advect large vorticity southward and result in positive RWS anomalies ($-V_\chi' \cdot \nabla\bar{\zeta}$, shown by purple contour) around 25°S by crossing the maximum absolute vorticity gradient latitude that is intense in the austral winter. Moreover, corresponding to the upper level convergence over southeastern Australia, descending motion in the troposphere induces vortex stretching in the upper troposphere and produces a pronounced patch of negative RWS anomalies ($-\bar{\zeta}D'$, green contour) over southeastern Australia and the central South Pacific. This subsidence over southeastern Australia results in vortex shrinking, which drives additional equatorward flow in the lower troposphere. A low-level high pressure anomaly stands out to its left to maintain the Sverdrup balance (Fig. 5.3) (Hoskins and Karoly 1981). The low-level southerly winds and

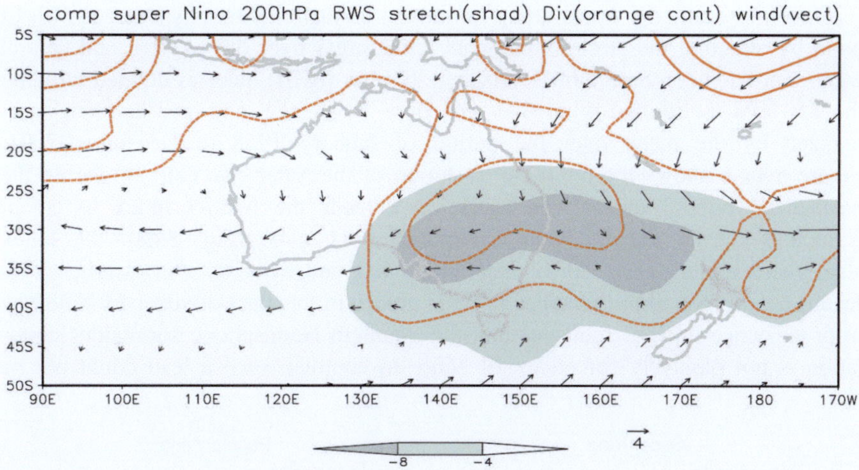

Fig. 5.2 Super El Niño composite of anomalous RWS stretching term (*shading*; 1e-11 1/s^2), winds (*vector*; m/s), and divergence (*orange contour*; contour interval: 5e + 5 s^{-1}) at 200 hPa during JJASON (Reprinted from Hong et al. 2014 with permission from John Wiley & Sons, Inc., New York, USA)

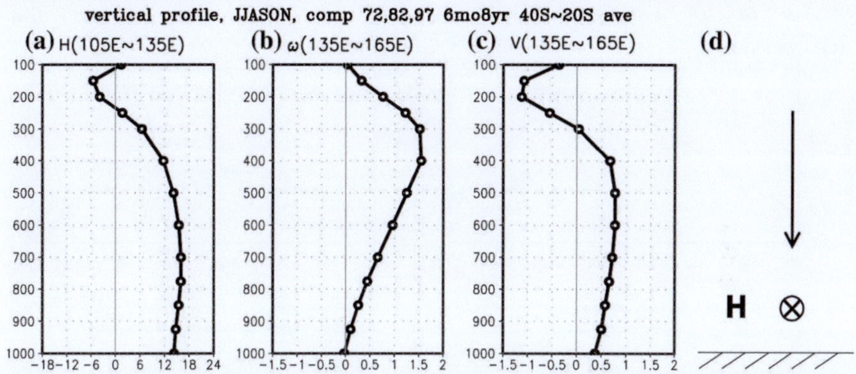

Fig. 5.3 Vertical profile of super El Niño composite for (**a**) geopotential height (m) averaged over 20°S–40°S, 105°E–135°E, (**b**) vertical pressure velocity (10^{-2} pa/s) averaged over 20°S–40°S, 135°E–165°E, and (**c**) meridional wind (m/s) averaged over 20°S–40°S, 135°E–165°E during JJASON. (**d**) Schematic diagram of Sverdrup balance in the Southern Hemisphere. *Arrow* denotes descending motion. Southerly wind is shown by *arrow pointed into the page*, and the letter *H* indicates high pressure (Reprinted from Hong et al. 2014 with permission from John Wiley & Sons, Inc., New York, USA)

the Australian high are thus integrated in a transverse cell to facilitate El Niño's growth. Individual El Niño events are provided in Figs. 3.15–3.16, which show that super El Niño events have similar spatial structure to the composite while regular El Niño events lack resemblance of super El Niño.

Since the southerly winds along northeastern Australia directly interact with El Niño by affecting the strength of zonal wind stress over the western and central equatorial Pacific, a Southern Hemisphere booster (SHB) index is defined utilizing the 850 hPa meridional wind averaged over 10°S–30°S and 140°E–170°E normalized by the mean over the period of January 1958 to December 2001. Figure 5.4a–b shows the temporal evolution of the transverse cell comparing the normalized Niño-3 index. The SHB index leads the Niño-3 index by about 3 months during the El Niño onset/developing stage (Fig. 5.4a); the index strengthens rapidly from April, peaks in August, and weakens after October. This feature is only found in the super El Niño group. In the class of super El Niño, the prior existence of SHB indicates that the Southern Hemispheric anomalous circulation is not passively forced by El Niño. In contrast, such a lead could not be

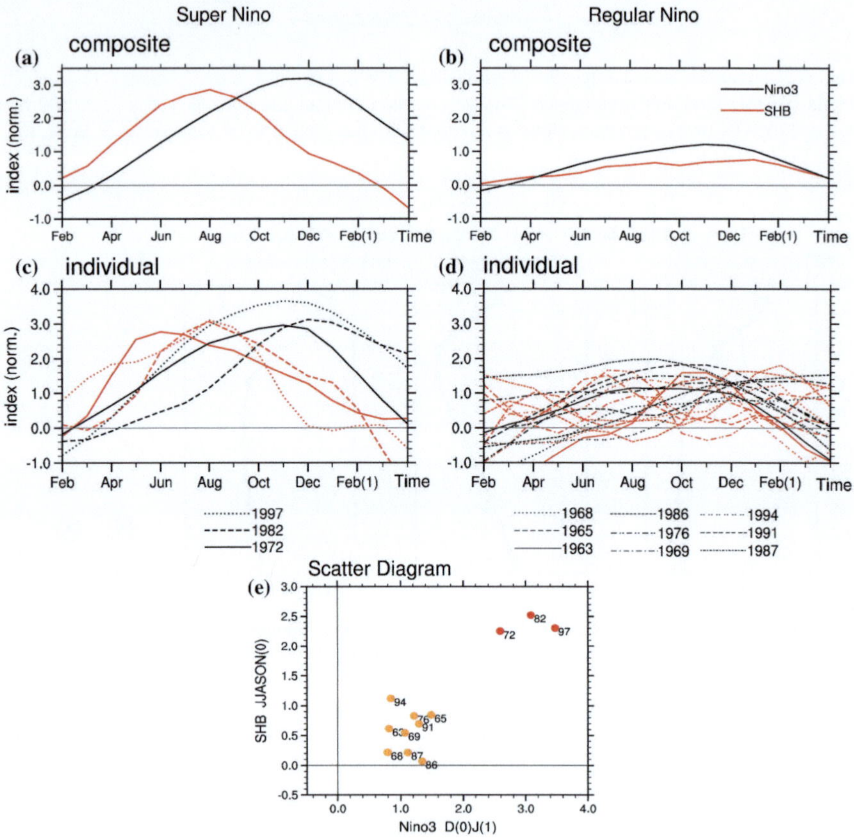

Fig. 5.4 (**a**) Normalized SHB index (*red*) and normalized Niño-3 index from February of the El Niño year to April of the following year for super El Niño composite. (**b**), (**c**), and (**d**) are the same as (**a**) except for regular El Niño composite, individual super El Niño, and individual regular El Niño, respectively. (**e**) Scatter diagram of normalized JJASON(0) mean SHB index against normalized D(0)J(1) mean Niño-3 index for the 12 El Niño events (Reprinted from Hong et al. 2014 with permission from John Wiley & Sons, Inc., New York, USA)

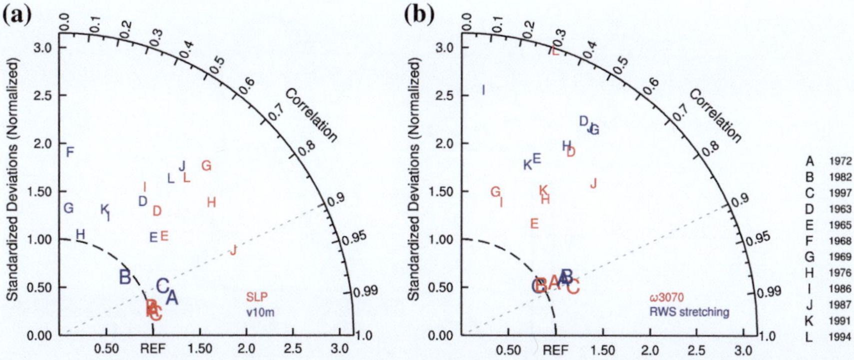

Fig. 5.5 Talyor diagram for spatial pattern of (**a**) SLP (*red*) and 10-m meridional winds (*blue*) and (**b**) vertical pressure velocity averaged over 300–700 hPa (*red*) and 200 hPa RWS stretching term (*blue*) between rescaled El Niño cases (*letters*) and rescaled super El Niño composite (*reference*). The *radial distance* from the origin is the standard deviation of an El Niño event normalized by that of super El Niño composite, and the *azimuthal angle* is the correlation between the spatial pattern of an El Niño case and the reference (Reprinted from Hong et al. 2014 with permission from John Wiley & Sons, Inc., New York, USA)

detected in regular El Niño (Fig. 5.4b), and the magnitude of the transverse cell is either too weak or absent altogether. The lack of SHB in regular El Niño is consistent with its sluggish growth and small amplitude. More details on indices the SHB index and Niño-3 index for individual El Niño events are provided in Fig. 5.4c–d, which displays temporal evolutions similar to the composite. Figure 5.4e contrasts the SHB index in JJASON(0) against the Niño-3 index in D(0) J(1) for the 12 El Niño events. The scatter diagram indicates that the strength of the transverse cell and the magnitude of El Niño are positively correlated (r = 0.87). The magnitudes of both for super El Niño events are proportionally larger than those for regular El Niño events. It appears that only El Niño events with the SHB index larger than two standard deviations during summer–autumn can grow into a super El Niño in the following winter.

One may suspect that regular El Niño events have a similar Southern Hemispheric structure but cannot be identified visually due to its weak strength. To examine this possibility, Taylor diagrams (Taylor 2001) are used or analyzing pattern similarity for the associated spatial fields, which are rescaled by the November–January mean Niño-3 SST anomaly for each El Niño case to mitigate the amplitude difference between the super and regular El Niño events to a large extent. The rescaled super El Niño composite is used as the reference. The SLP (50° S–15°N, 90°E–90°W), vertical pressure velocity averaged over 300–700 hPa (35° S–15°N, 90°E–120°W), 10 m meridional winds (30°S–10°N, 120°E–180°), and the 200 hPa RWS vortex stretching term (45°S–15°S, 70°E–90°W) are chosen as the four representative fields of the transverse cell to quantify spatial similarity of each "normalized" El Niño case. As shown in Fig. 5.5a–b, the three super El Niño events are clustered together and strikingly distinct from the rest. Their standard deviations

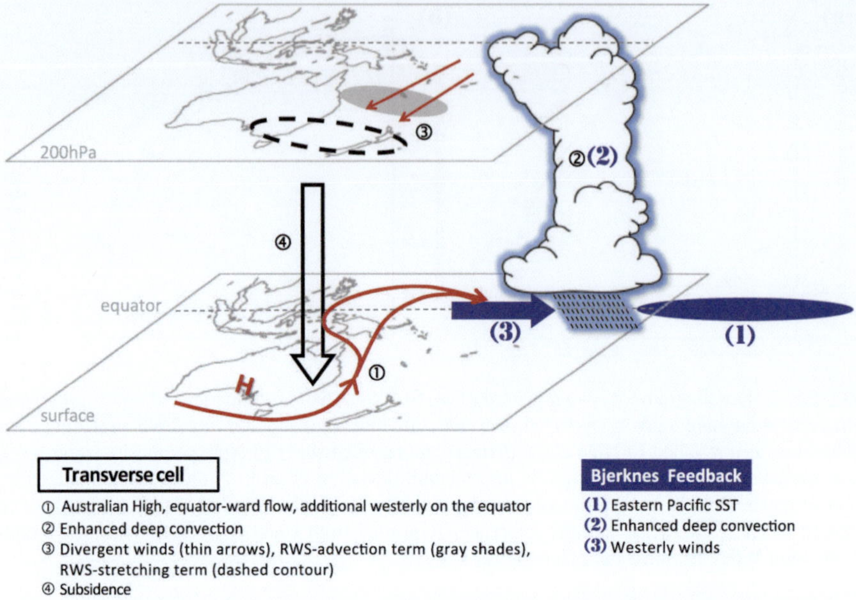

Fig. 5.6 Schematic diagram of super El Niño development illustrating how a transverse cell with main features in the Southern Hemisphere (marked by *numbers in open circle*) interacts with the Bjerknes feedback regime in the central equatorial Pacific (depicted by *large blue numbers in parentheses*) (Reprinted from Hong et al. 2014 with permission from John Wiley & Sons, Inc., New York, USA)

are close to one, indicating that the magnitude of the spatial pattern is also close to the reference point (their composite). In contrast, regular El Niño events spread out in the Taylor diagrams due to the lack of resemblance to super El Niño. The large spatial standard deviation of regular El Niño events results from the lack of coherence in the transverse cell.

Figure 5.6 presents a schematic diagram demonstrating the extraordinary amplification of super El Niño events through a Southern Hemispheric transverse cell during the developing stage. Although the growth of super El Niño is mainly caused by the Bjerknes feedback in the equatorial region, the transverse cell further strengthens the low-level tropical westerly winds with the aid of the Australian high to booster super El Niño's growth during austral winter.

References

Annamalai, H., S. Kida, and J. Hafner, 2010: Potential Impact of the Tropical Indian Ocean–Indonesian Seas on El Niño Characteristics*. *J. Clim.*, **23,** 3933-3952.

Hong, L.-C., LinHo, and F.-F. Jin, 2014: A Southern Hemisphere booster of super El Niño. *Geophys. Res. Lett.*, 2014GL059370.

Hoskins, B. J., and D. J. Karoly, 1981: The steady linear response of a spherical atmosphere to thermal and orographic forcing. *J. Atmos. Sci.*, **38,** 1179-1196.

Rodwell, M. J., and B. J. Hoskins, 1995: A Model of the Asian Summer Monsoon.Part II: Cross-Equatorial Flow and PV Behavior. *J. Atmos. Sci.*, **52,** 1341-1356.

Sardeshmukh, P. D., and B. J. Hoskins, 1988: The generation of global rotational flow by steady idealized tropical divergence. *J. Atmos. Sci.*, **45,** 1228-1251.

Stephens, D. J., M. J. Meuleners, H. van Loon, M. H. Lamond, and N. P. Telcik, 2007: Differences in Atmospheric Circulation between the Development of Weak and Strong Warm Events in the Southern Oscillation. *J. Clim.*, **20,** 2191-2209.

Taylor, K. E., 2001: Summarizing multiple aspects of model performance in a single diagram. *J. Geophys. Res.*, **106,** 7183-7192.

Chapter 6
Preliminary Result of Model Analysis

Within the framework of the CMIP3 and CMIP5, several recent papers (e.g., Lengaigne and Vecchi 2010; Santoso et al. 2013; Cai et al. 2014) have laid the foreground for looking into various aspects of super (extreme) El Niño events. Here, preliminary results from the output of the GFDL-ESM2M, generally regarded as one of the most reliable models to reproduce El Niño signals (see Supplementary Figs. 8 and 11 in Cai et al. 2014) are shown in this chapter.

6.1 Super El Niño Identification

To be consistent with Fig. 1.1, a super El Niño in the model is simply defined by the normalized Niño-3 SST greater than 2.5 standard deviations averaged over November–December–January, and a regular El Niño, by the value greater than 0.5 but less than or equal to 2.5 standard deviations. A total of 12 super and 109 regular El Niño events are found over a 500-year simulation period of the preindustrial experiment in the GFDL-ESM2M output.

6.2 Super El Niño Behaviors

The simulated super El Niño composite and simulated regular El Niño composite plotted the same as Fig. 5.1 are shown in Fig. 6.1. Although the GFDL-ESM2M is able to simulate super El Niño events, the temporal evolutions of super El Niño in the model are different from those the observation (Fig. 6.1e). The simulated super El Niño events have a significantly longer duration due to earlier onset and later termination and have double peaks in August (year 0) and February (year 1). The inconsistent behavior before a super El Niño rapidly develops implies that the simulated super El Niño may not appropriate to be utilized to evaluate the proposed preconditions of super El Niño in the observation. Thus, the preconditions of simulated super El Niño are not shown here. However, the super El Niño events in the model feature a strong SHB signal during the developing stage, similar to what

L.-C. Hong, *Super El Niño*, Springer Theses,
DOI 10.1007/978-981-10-0527-5_6

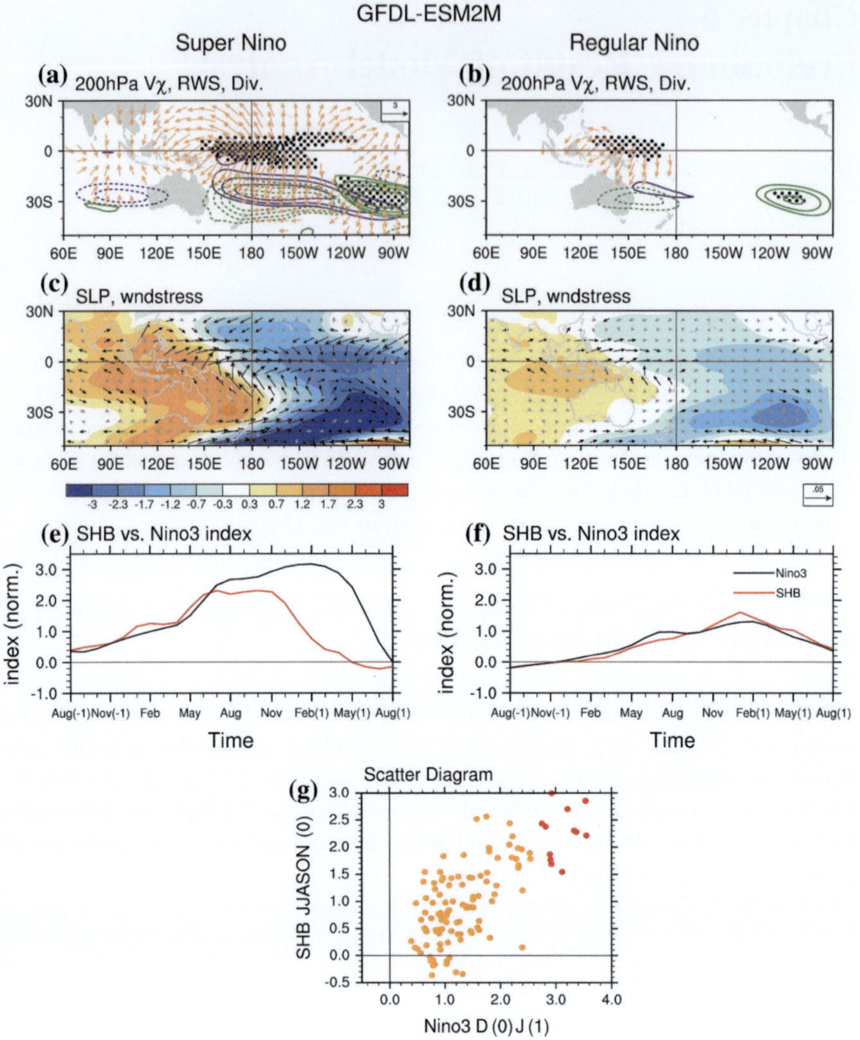

Fig. 6.1 **a–d** Same as Fig. 5.1 except for GFDL-ESM2M and the following variables: the vertical motion in Fig. 5.1a is replaced by 200-hPa divergence (only value >1.25e-6 1/s is shown); 10-m wind in Fig. 5.1c is replaced by wind stress (magnitude >0.01 pa is plotted in *black*). **e** and **f** Same as Fig. 5.4**a–b** except for the time span expands from August before the El Niño year to August of the following year. **g** Same as Fig. 5.4**g** (Reprinted from Hong et al. 2014 with permission from John Wiley & Sons, Inc., New York, USA)

is found in the observation. For the regular El Niño events in the model, the SHB signal is much weaker. Moreover, it can be inferred that the effect of SHB in terms of its role in strengthening El Niño in the sense that the SHB tends to peak ahead of the Niño-3 index as found in the observation. There exist discrepancies, however, between model simulation and observations. The Australian high in the model is further to the northeast, so the domain for evaluating the SHB index has to be shifted to 20°S–0° and 150°E–180°. Furthermore, the associated SHB in the model also shows a longer duration. These biased features of simulated El Niño are likely due to model biases in simulating climate mean state and annual cycle.

References

Cai, W., and Coauthors, 2014: Increasing frequency of extreme El Nino events due to greenhouse warming. *Nature Clim. Change*, **4,** 111-116.

Lengaigne, M., and G. Vecchi, 2010: Contrasting the termination of moderate and extreme El Niño events in coupled general circulation models. *Clim Dyn*, **35,** 299-313.

Santoso, A., and Coauthors, 2013: Late-twentieth-century emergence of the El Nino propagation asymmetry and future projections. *Nature*, **504,** 126-130.

Chapter 7
Conclusions and Future Work

7.1 Conclusions

This study was set out to explore the unique features, preconditions, and causes of a newly proposed El Niño group, super El Niño. Three main issues of super El Niño events are addressed:

1. Since super El Niño events are selected using a threshold of magnitude, whether they have enough common and unique features to group a new class besides their magnitude.
2. Can we predict the occurrence of a super El Niño? When does the earliest signal emerge and how to trigger a super El Niño?
3. How do super El Niño events attain such strong amplitude? What is the underlying mechanism to boost their growth?

First, the grouping of super El Niño is confirmed by the *K*-means clustering analysis, which shows that super El Niño events are classified as a distinct group from other ENSO events according to the primary features of ENSO during their mature stage. Furthermore, by comparing 12 El Niño events covered by ERA-40 data, super El Niño events exhibit distinct spatial and temporal structure, such as the PMM and the SH transverse cell during their evolution. The unique Southern Hemispheric anomalous pattern is further supported by Taylor diagrams after rescaling to mitigate the amplitude difference between the super and regular El Niño events. Some unique signatures or signs can even be traced back to at least 2 years before super El Niño events, and the modulation of super El Niño events on the tropical Pacific may last for 2 years, thus boosting ENSO predictability tremendously. By demonstrating these unique features, it is suggested that super El Niño events should be categorized as a new group.

Among these features, the super-ma La Niña and the Hawaii and Philippine Sea low are hypothesized as two necessary preconditions for super El Niño onset. Super-ma La Niña recharges the upper-ocean heat content along the equatorial

L.-C. Hong, *Super El Niño*, Springer Theses,
DOI 10.1007/978-981-10-0527-5_7

Pacific in the spring of the El Niño year in a special way. It stashes anomalous subsurface warm water in the tropical western Pacific in the year before a super El Niño event, and the subsurface warm water gradually converges to the western equatorial Pacific and contributes to the SST warming in the eastern equatorial Pacific (in April–May). In addition, the mild cold SST anomaly over the central tropical Pacific during a super-ma La Niña event facilitates SST warming over this region by the following precondition, the Hawaii low and the Philippine Sea low.

In the winter before a super El Niño event, the Hawaii low sprawls toward to the central equatorial Pacific via the thermodynamic mechanism, the WES feedback, and the dynamical mechanism, the Sverdrup balance, and then merged with the Philippine Sea low. The sprawling of the Hawaii low generates a pronounced PMM pattern in the north tropical Pacific. During the same period, the merged large-scale low anomaly effectively warms the central equatorial Pacific by low-level westerly winds. As a result, the whole equatorial Pacific SST turn positive in April–May, and in the northern-central equatorial Pacific, vigorous convection is produced due to the warm SST anomaly, indicating the onset of a Super El Niño.

The northern-central equatorial Pacific is a key box region for super El Niño events, since the enhanced convection in this region initiates the SH transverse cell, which boosts the growth of these El Niño events. The SH transverse cell mainly consists of an anomalous surface high in southern Australia, enhanced convection over the ENSO regime in the central equatorial Pacific, and a pattern of upper-level Rossby wave source in between. This transverse cell is pronounced in the super El Niño development phase during the boreal summer/austral winter. It is featured with the lower-level equatorward wind anomalies along northeast Australia, which can bring southern cold surge thus more westerly wind burst events to strength the El Niño. Thus, this cell is hypothesized to serve as a booster to facilitate the runaway growth of a super El Niño. The output of CMIP5 GFDL-ESM2M supports the results based on the observation, despite some discrepancies between the model and the observation.

In summary, to breed a super El Niño, assistance outside the tropical Pacific are necessary. With the aid of the super-ma La Niña, which may be modulated by the negative IOD events, and the large-scale low SLP anomaly in the north Pacific in the previous winter, a super El Niño is triggered. Moreover, the anomalous circulation from the Southern Hemispheric during its developing stage further accelerate and boost its growth, thus resulting in a super El Niño.

7.2 Future Work

1. Our major goal is to modify the core parts of ENSO theory, namely, Bjerknes instability and phase-turning theory (the recharge-discharge oscillator, RDO), especially for those super El Niño cases. We propose that the middle latitude feedback contributes significantly for El Niño's triggering, developing and

phase turning. We intend to measure the non-Bjerknes effect by a quantified instability index. A monsoon oscillator (maybe in a supplementary way) is very much in mind.

2. For ocean process we challenge the current meridional symmetry of the RDO. We want to make an extra effort to incorporate the equatorial circulation into the ENSO scenario, especially the local branch of shallow subtropical convection cell in lieu of the western North Pacific monsoon. In other words, the seasonality of extra-tropical circulation will be scrutinized closely.
3. It is believed that a super El Niño case carries much earlier sign before its onset; thus it offers extended predictability. A super El Niño episode also has a good command of phase turning; it sharply transformed into a La Niña in the following year. In all, we are interested a stretch of 3–7-year deterministic cycle wrapped around a super El Niño peak that may partially explain why this special type of El Niño appears within a decadal modulation. Modeling effort is eagerly pursued. Some side issues like "Why the 1972 El Niño is regarded a super El Niño event?", "Is 2014 a super El Niño year?" are currently under study.

Appendix

A.1 Sverdrup Adjustment

The Sverdrup vorticity balance of the steady flow is

$$\beta v \approx -f\left(\frac{\partial u}{\partial x} + \frac{\partial v}{\partial y}\right), \tag{A.1}$$

where f is the planetary vorticity, $\beta \equiv (\mathrm{d}f/\mathrm{d}y)$is the planetary vorticity gradient, and $\left(\frac{\partial u}{\partial x} + \frac{\partial v}{\partial y}\right)$ is the horizontal divergence. If we consider this balance at the 850 hPa pressure level for anomalous flow (with respect to the seasonal climatology), Eq. (A.1) may be written as

$$-\left(\frac{\partial u'}{\partial x} + \frac{\partial v'}{\partial y}\right)_{850\,\mathrm{hPa}} = \frac{\beta v'_{850\,\mathrm{hPa}}}{f}. \tag{A.2}$$

In tropics, the low-level convergence is highly correlated with the local tropospheric upward motion. Thus the 850 hPa convergence anomaly of Eq. (A.2) may be rewritten in terms of vertical pressure velocity above the 850 hPa:

$$\omega'_{\mathrm{above}\ 850\,\mathrm{hPa}} \approx a^{*}\frac{\beta v'_{850\,\mathrm{hPa}}}{f}, \tag{A.3}$$

where a is assumed as a constant which can be obtained from the linear regression for simplicity. Here, we use vertical pressure velocity averaged from 700 to 300 hPa to regress on the 850 hPa convergence, and a is -2.22×10^4 Pa over the tropical North Pacific Ocean for the period of 1958–2001. Equation (A.3) expresses the meridional planetary advection by anomalous pole-ward (equator-ward) winds balanced with the anomalous upward (downward) motion in the troposphere. Since $f = 0$ at the equator, we only calculate the Sverdrup adjustment outside 5°S–5°N.

L.-C. Hong, *Super El Niño*, Springer Theses,
DOI 10.1007/978-981-10-0527-5

A.2 Oceanic Mixed Layer Heat Budget

The anomalous temperature equation for the mixed layer may be written as (Li et al. 2002)

$$\frac{\partial T'}{\partial t} = -(V' \cdot \nabla \overline{T}) - (\overline{\mathbf{V}} \cdot \nabla T') - (\mathbf{V}' \cdot \nabla T) + \frac{Q}{\rho C_P H} + R \tag{A.4}$$

where T represents the mixed layer temperature; $\mathbf{V}$ is three-dimensional (3D) ocean current, which is defined as the vertical average from surface to the bottom of mixed layer; ∇ denotes 3D gradient operator; $(\ )'$ represents the interannual anomaly variables; $\overline{(\ )}$ represents the monthly-mean climatological variables; $-(\mathbf{V}' \cdot \nabla \overline{T})and-(\overline{\mathbf{V}} \cdot \nabla T')$are the linear advection terms associated with anomalous current and anomalous temperature, respectively; $-(\mathbf{V}' \cdot \nabla T')$ denotes 3D nonlinear temperature advection term; Q is the net surface heat flux including shortwave radiation, long wave radiation, surface latent, and sensible heat fluxes; a positive heat flux indicates heating the ocean.; R represents the residual term; $\rho = 10^3$ kg m^{-3} is the density of water; $C_P = 4000$ J kg K^{-1} is the specific heat of water; and H denotes the mixed layer depth and in this study H is chosen to be 45 m. Although 45 m is an appropriate choice for the eastern equatorial Pacific, the climatological value of the mixed layer depth in the north-central equatorial Pacific (0°–7.5°N, 155°E–170°W) is deeper (about 65 m). However, the result remains similar when this method is applied over the north-central equatorial Pacific using $H = 65$ m. Thus the analysis is not sensitive to H in the region concerned. Note that the vertical temperature gradient in the eastern equatorial Pacific is calculated as the temperature difference between the thermocline (about 65 m) and the mixed layer to correctly quantify the vertical advection since the anomalous subsurface temperature maximum is at the thermocline rather than at the bottom of the mixed layer. Since GECCO only provides net surface heat flux, surface radiation, surface heat flux data of ERA-40 were used to evaluate the relative importance for SST warming. Moreover, surface latent heat flux of ERA-40 was used to decompose into the components associated with wind speed and surface to near-surface moisture difference based on Chiodi and Harrison (2007).

MIX
Papier aus verantwortungsvollen Quellen
Paper from responsible sources
FSC® C105338

If you have any concerns about our products,
you can contact us on
ProductSafety@springernature.com

In case Publisher is established outside the EU,
the EU authorized representative is:
Springer Nature Customer Service Center GmbH
Europaplatz 3, 69115 Heidelberg, Germany

Printed by Libri Plureos GmbH
in Hamburg, Germany